MW01041792

Engineering Perfection

Revolutionary Bioethics

Series Editor: Rachel Haliburton, University of Sudbury

Revolutionary Bioethics is a new series composed of scholarly monographs and edited collections organized around specific topics that explore bioethical theory and practice through the frameworks provided by feminist ethics, narrative ethics, and virtue ethics, challenging the assumptions of mainstream bioethics in the process. Contemporary mainstream bioethics has become ideological and repetitive, a defender of activities that bioethics was originally created to critique, and an apologist for unethical practices and policies in medicine that it once saw itself as fighting against. Taking its title from recent work being done on MacIntyre's neo-Aristotelian ethics, *Revolutionary Bioethics* is organized around the idea that bioethics needs to reform both its theory and practice, and its goal is to begin the conversation about what a transformed bioethics—one that is unafraid to explore new theoretical approaches, and to examine and critique current bioethical practices—might look like.

Titles in the series

Engineering Perfection: Solidarity, Disability, and Well-being, by Elyse Purcell

Physician-Assisted Suicide and Euthanasia: Before, During, and After the Third Reich, Edited by Sheldon Rubenfeld and Daniel P. Sulmasy, with Astrid Ley

Engineering Perfection

Solidarity, Disability, and Well-being

Elyse Purcell

LEXINGTON BOOKS
Lanham • Boulder • New York • London

Published by Lexington Books
An imprint of The Rowman & Littlefield Publishing Group, Inc.
4501 Forbes Boulevard, Suite 200, Lanham, Maryland 20706
www.rowman.com

British Library Cataloguing in Publication Information Available

Library of Congress Cataloging-in-Publication Data on File

ISBN 978-1-7936-2411-6 (cloth | alk. paper)
ISBN 978-1-7936-2412-3 (electronic)

™ The paper used in this publication meets the minimum requirements of American National Standard for Information Sciences—Permanence of Paper for Printed Library Materials, ANSI/NISO Z39.48-1992.

For Sebastian and Kyle

Contents

Acknowledgments

The research in this book was developed over several years. I am much indebted to my colleagues at SUNY-Oneonta for their sound advice and for providing me with the institutional support for this project. I would also like to thank Elizabeth Dunn, Dean of the Arts and Humanities, who provided funding for the research of this project. Additionally, I would like to thank my dissertation advisors from Boston College and my students in my bioethics classes for pushing me on topics covered in this work.

I would like to thank John Abbarno and the members of the American Society for Value Inquiry for their valuable feedback on the central ideas of this book. Second, I would like to thank Matthew Brown and Magdalena Grohman and the participants of the annual Center for Values in Medicine, Science, and Technology (VMST) conference at the University of Texas at Dallas for fostering the development of this project. Many of the positions in this book were inspired by the stimulating discussions at the VMST from 2014 to 2020. Third, I would like to thank the organizers and participants at the following conferences and institutes for their feedback and encouragement from 2016 to 2020: the Radical Philosophy Association, the Institute of Cross Cultural Studies and Academic Exchange, the Society for Indian Philosophy and Religion, the International Health Humanities Conference, the West Coast Methods Institute, the Society for Ricoeur Studies, the Diversity Institute and the Center for Ethics, Peace and Social Justice at SUNY-Cortland.

I wish a special thank you to Robin Smith, who advised me to not let other things interfere with my research. Additionally, I would also like to thank the executive committee members of the American Philosophical Association—Central Division from 2016 to 2021 for their encouragement in my development as a scholar. Much appreciation is also deserved for Amy Ferrer,

Cheshire Calhoun, Lanier Anderson, Sven Bernecker, Becko Copenhaver, Jeff Dunn, Jeremy Cushing, Dom Lopes, and Ben Caplan for their mentorship and feedback.

I would like to thank my parents, my in-laws, and friends for their support and patience: the challenging holidays, the delayed phone calls, and the disrupted family vacations. Thank you, Jeff and Nora, for your encouragement and for sharing with me perspectives on these topics from Mexico. Thank you, mom and dad, for listening to my philosophical ideas, discussing them with your friends and colleagues, and for even trying to follow along in the news on genetic technologies.

Most of all, I would like to thank my husband and fellow philosopher, Sebastian, who read several drafts of this manuscript and offered irreplaceable advice, who discussed the ideas and arguments with me during many evenings, and who listened to my frustrations. Thank you for not giving up on me when I had given up on myself. You have offered more than one could ask from a loving partner and friend.

Permissions have been obtained for the following. Kyle Lawson's story and quotations are printed with permission. Reprinted material from Purcell, Elizabeth (2014), "Oppression's Three New Faces: Rethinking Iris Young's 'Five Faces of Oppression' for Disability Theory," in *Diversity, Social Justice and Inclusive Excellence: Transdisciplinary and Global Perspectives*, Mecke Nagel and Seth Asumah (ed.), SUNY Press. Reprinted with permission. Reprinted material from Purcell, E. B. (2016). "Ethics and Mental Illness: An Intercultural Approach," *Wagadu: A Journal of Transnational Women's and Gender Studies*, Special Issue: Epistemic Injustice in Practice, vol. 15, 115–38. Reprinted with permission. Tia Chivel's name, their family's names, and their story is printed with permission.

Introduction

GENETIC TECHNOLOGIES AND MY EXISTENTIAL JOURNEY

What do I owe my future child? This has been a question on my mind for some time. The research for this book took place during a time of personal crisis in my life: the problem of infertility that my spouse and I faced. Working with a fertility clinic and going through the adoption process wove together three strands in my mind, which form a conceptual shape for philosophers who discuss the topic of disability: vulnerability, embodiment, and capitalism.

These personal experiences influenced the philosophical questions that came to my mind. The questions concerned how capitalism alters the real and concrete conditions for how we should treat individuals within our society and what obligations social institutions have to members of that society. Thus, the questions that came to my mind were how does capitalism distort what we owe each other? And even closer to my heart, how does capitalism distort what I owe my future child? Perhaps it is easier to explain with a few stories.

In the process of applying for adoption, my partner and I were asked to complete various forms, provide financial statements, and to have our physicians attest to our health "fitness." Once we were provisionally approved, we had numerous discussions with our case worker and also completed a home study. What surprised me were some of the questions our case worker asked us. Printed on a questionnaire were the following possible inherited disabilities carried by the expected parent or extended family: bipolar disorder, depression, schizophrenia, intellectual disability, and autism. We were instructed to write "yes," "no," or "willing to discuss." Being supportive, the case worker and the adoption agency left instructions to educate ourselves and offered training on raising a child who might have one of these inherited disabilities.

Because I have a mental health condition, the adoption process forced me to grapple with my mixed emotions about being a parent and to realize the obligations one takes on when one becomes a parent. Could I be good enough to parent a child with a disability? The answer to this question after much reflection and education was "yes." Regardless of having had a successful fertility treatment or choosing to adopt, the possibility of having a child with a disability has always been present for potential parents.

Our decision to pursue adoption came after several unsuccessful fertility treatments and being diagnosed as "infertile" for unknown reasons. After my last treatment, I met with a nurse for my consultation on in-vitro fertilization (IVF). The nurse explained to me the procedure, and then casually remarked, "if your eggs have defects, then we will just use a donor egg." She then told me she was unsure how much we had left on insurance, so she would set up a meeting with financing. In an intense moment of grief, I left the clinic distraught, outraged, and decided not to return.

This moment of grief, though, became later a moment of enlightenment. My professional experience as a Marxist bioethicist reframed my personal interaction with the nurse. In a capitalist society, the nurse was doing what she had been trained to do. Her job was to ensure the "best" product, and if my egg had a "defect," then she would not be offering me the "best" product available. The future child my partner and I sought after to complete our family was only a commodity to be sold to another desperate couple. As I continued my research as a philosopher, another question came to mind. This question was about the future. What if the nurse had instead asked me "would you want to enhance your egg? Or would you like an enhanced egg?" And of course, there would be an additional price for the "upgrade."

Since the development of genetic engineering tools, such as CRISPR-Cas9, we now have the possibility to make the nurse's hypothetical questions a reality. Philosophers who endorse genetic enhancements, including those who identify as transhumanists, would welcome this possibility. In theory, a potential parent could choose to select or enhance traits for their future child. One of the central issues surrounding fertility treatments and enhancement has turned on choosing a child with a disability.

Many philosophers have argued that up until now having a child with a disability was left to chance or luck. With the advent of these developing technologies parents would be able to choose against having a child with a disability. What fertility treatments and adoption has taught me, however, is that many of my choices over the process of becoming a parent were still and foreseeably outside of my control.

One's actual choice is and will be limited for several reasons. First, when a potential parent wants to have a future child, they experience societal and cultural pressures to have a particular kind of child. These factors often influence their decisions and can sometimes undermine the parents' reproductive autonomy so that their choice is not completely free. Second, pregnancy does not happen in an incubator because women are not fetal containers. Philosophers in favor of genetic enhancement have assumed that enhanced children will not have complications which can occur in pregnancy and in labor. Yet, these complications can alter the outcome of what one has "chosen" to enhance. Third, becoming pregnant cannot be literally controlled. Many men and women seek fertility treatments because pregnancy has been difficult for them. Some, by contrast, become pregnant without the intention because contraception failed. For others, pregnancy proves successful only the first time by luck. In summary, there are significant factors beyond genetic selection that determine the traits of a child, and we cannot control these. What one can control, however, is how one chooses to be a parent. One becomes a parent by chance; one fulfills one's obligations by choice.

I believe, though, that the issue of choosing a child without a disability points to a second motivation behind the enhancement proponents' theories: a fear of our own fragility. While philosophers in the enhancement debate have been divided on the use of these technologies, they have yet to address human living in particular circumstances. These philosophers view genetic technologies as tools to protect humans from being vulnerable. I think this is a mistake.

Part of the research for this book was inspired by a close friend of mine, Kyle Lawson. Kyle has a rare form of Juvenile idiopathic arthritis. Since about the age of seven, his symptoms have included joint swelling and destruction, stiffness, fatigue, and pain. Kyle's case is quite rare, because it affects every joint in his body. The particular effects of his condition are so rare that medical doctors in his community could not offer a prognosis on its progression. As Kyle has said to me many times, "I have no pain free moments, just variably intense ones." Although his cognitive tests have verified his significant intellectual gifts, because his condition limited and damaged his growth, Kyle was discriminated against in college by his peers. The assistance he needed was not available, and his professors were not equipped to meet his needs. One college peer even harassed him. As a result of these incidents, Kyle was unable to complete his university program. Forces such as these continued to exclude him from a normative system not sensitive for those like him.

Had genetic technologies been around for Kyle's parents, they might have been advised to terminate the pregnancy of Kyle. Would that have been the morally right advice? Would it have been better that Kyle had never been born? What gives us the normative grounds to conclude that decision?

I suspect that these philosophers seek to protect themselves from a life like Kyle's. A life they would deem one not worth living. Genetic technologies promise invulnerability. Science promises to offer protection from the five areas of vulnerability that they fear: illness, aging, death, misfortune, and suffering.

Vulnerability, however, is part of human living, and our vulnerability is not a facet of life that should be feared. There is always the chance that one will get hurt, that one will lose someone, and that one will have personal struggles. Living with a disability can add a dimension to one's perspective about what is worthwhile in life.

Like many of us, Kyle has endeavored to find meaning in life and love. Since leaving the university, he found a significant romantic partner, became an ardent student of philosophy and the liberal arts, an activist on behalf of persons with disabilities, a leading counselor for an arthritis support chapter, and a yoga instructor where he puts some of his philosophy into practice. Kyle's case may give one pause about the control one has over one's future child's life. He himself has told me that his disability has given him a personal depth and a richness of human experience that he would not have had without it. For Kyle, his birth and life, even with the daily pain, has been good overall. As he has said, while he does "objectively feel more physical pain than most, subjectively" he knows that he "suffers less than most."

Some philosophers may argue that Kyle's case is a tragic one, and that one of the aims of genetic technologies is to prevent suffering and tragedy. But what constitutes a tragic life? Suffering runs on a continuum. Where can we pinpoint how much is too much? The truth is much of our avoidance of suffering and tragedy in life is bald luck. Even if we make a choice using a genetic enhancement, that choice does not grant one control over how well-lived the future child's life will be. Kyle has experienced friendship, romantic love, and family ties. He has appreciated fine art, music, world philosophies, and the physical activity of yoga. Are these not some of the components on objective list theories of well-being? Just because he has experienced pain, discrimination, and stigma, it does not mean that he has a life not worth living.

After my partner and I decided to choose adoption, I took time to reflect on the emotions I experienced. Sometimes I doubted my parenting abilities. Other times I wondered what the future child we might have might be like. I also thought about the world in which we lived. I asked myself, "what would

I want for my future child?" I realized that I did not care whether the child had an excellent memory, was good in athletics, or was extraordinarily beautiful. I only wished that they would be happy, could pursue their interests, and know that they were loved. I wanted them to love themselves and realize their gifts. I wanted them to have friends and romantic interests. And finally, I thought, I would fight anyone who ever said my child did not belong in our society.

This fear of vulnerability and the motivations of the capitalist class to accumulate wealth as I witnessed at the fertility clinic converge on the topic of genetic enhancement. Genetic technologies have now provided scientists with the medical tools to change particular inherited traits. How we use these technologies, however, is guided by a normative dimension. The choice before us is how we will use them to do good. Without considering the societal and environmental challenges of our present world, how can we hope that these technologies will bring the increase in well-being and societal justice that we desire?

This book aims to answer this question. As we have witnessed in the global pandemic of 2020, "moral" people have come to different conclusions about what is best for their children's education and health. Those with financial and social resources hired private tutors or created "pods" for learning in small cohorts. Others, without those resources, became dependent upon the state to provide a safe education for their children. The enhancement of certain moral traits cannot help parents make "better decisions" when the possibilities for those choices have been curtailed by societal factors that include a lack of resources and unchecked oppression.

Genetic technologies challenge us to reflect on the limited control we have over the well-being of a future child's life. Further, these technologies force us to face our fear of our own vulnerability. Genetic technologies are tools, but they will not be able to change the facets of luck and fragility that are part of the human condition. It is in this sense, that genetic technologies challenge us to reconceive of the good: to reject our conception of the good as human perfection, and instead, to reconceive of the good as differentiated solidarity. This reconception is what I call the Solidarity view, which begins with a reconception of the good as one's self-realization through the interdependent mutual recognition and co-belonging with others.

The Solidarity view holds that the endeavor of genetic enhancement is fundamentally misguided in two ways. First, genetic enhancements are possible products marketed to those who have fears about embodiment and vulnerability. In this sense, it is much like the commercialization of beauty products. Second, those who endorse enhancement hold to a liberal of conception society. My suspicion is that some fear being pushed out of the labor market because the bodies of those who are not enhanced will not be as exploitable in the labor market as those who are enhanced. Enhancement, then, not only proves to be a threat for those with nonconforming bodies; it is also a threat to those who do not "opt in" to the enhancement rat race. The goal instead, I argue, should be to abandon the liberal approach to society, and instead to reconceive of society through the lens of solidarity.

Within a liberal conception of society, capitalism operates at two levels which have been overlooked concerning enhancement. At the individual level, capitalism has a long history of exploiting our fears, such as a fear of illness, death, or pain, in order to capitalize on them. By selling us a product that could possibly alleviate those fears, the consumer easily makes the purchase. The fear of having a vulnerable child is what is central to the debate. And if capitalism can offer a product to alleviate this fear through genetic technologies, as the nurse offered me a donor egg, then its logic dictates that it will.

Yet, capitalism also operates on the systemic level. As a system, capitalism wants exploitable bodies. It is in this way that genetic enhancement provides a solution: enhanced workers who provide a larger surplus value. As a boundary social category, disability is interpreted differently from a capitalist viewpoint: how exploitable is the body? To clarify, I think that Marta Russell's conception of disability as a socially created category, which is derived from labor relations, is helpful. For Russell, disability is "a product of the exploitative economic structure of a capitalist society: one which creates and (then oppresses) the so-called disabled body as one of the conditions that allow the capitalist class to accumulate wealth" (2019, 2). Marterialist praxis requires that genetic technologies and the commodification of bodies in a capitalist system be viewed with suspicion.

These points bring together the approach of this book. Often, there is a preference for philosophical work to be composed in the abstract. And yet, it is through my existential journey that I came to these philosophical questions. Throughout this work, then, I will consider these questions from my personal perspective as well as draw from the testimony of people with disabilities (Kittay *et al.* 2001). Part of my methodology is to "think with" others, rather than independently, to develop my own thought. It is both performatively consistent and necessary, given the need to rely on others for testimony.

STRUCTURE AND CHAPTER OUTLINE

The philosophical debates concerning the use of genetic technologies often touch on the topic of human perfection (Bostrom 2005a).[1] Perfectionism as an ideal, within the tradition of Western philosophy and religion, considers capacities which perfect human nature to constitute well-being (Roduit, Baumann, and Heilinger 2013; Agar 2007).[2] And yet, the pursuit of perfection is contentious because of its sordid history. The twentieth century eugenics movement aimed to "improve the human gene pool" through biased and poorly understood genetic practices and experiments (Wikler 1999). The troubled history of the eugenics movement leads us to consider how social and normative values permeate our contemporary medical and scientific endeavors (Sparrow 2014b).[3]

The appeal of genetic technologies, I think, is that they offer a kind of physical and psychological invulnerability. They hope to prevent or postpone the dimensions of human existence in which we are vulnerable: illness, aging, death, misfortune, and suffering. These five areas might be called the *five faces of vulnerability*, which affect the existential dimension of our lives. The five faces target the "who" of our narrative identity and our sense of self within our lived experience (cf. Ricoeur 1992). Changing the material of our bodies will not necessarily change the facets of the human condition (cf. DeGrazia 2000, 2005).

The structure of this work follows the five faces of vulnerability. Throughout the work, I will consider how capitalism exploits and distorts our fears for ourselves and our future children within the discussion concerning genetic technologies. To do so, I will turn to personal stories and testimonies from those who have experienced the five faces and share what their experiences have taught me. From these testimonies, I bring together the central features of the Solidarity view. I spell out the components of this view in the following chapters.

Chapter 1 introduces the five faces of vulnerability and outlines the components of what I call the Solidarity view. In this chapter, I address our central fear of vulnerability, which is exploited and manipulated by capitalism. It does so through the myth of control. The thought is that if we control our bodies (e.g., by eliminating particular traits), then it is possible to protect ourselves from vulnerability.[4] And yet, the faces affect the existential and concrete dimensions of human living insofar as they alienate us from one another and from our lived experience. In this way, I think the debate concerning these technologies tends to miss its mark.

I argue that genetic technologies, such as enhancement, only serve as a temporary salve for the individual fears. Instead of changing our bodies, we need to reconceive of our societal practices. I propose that this reconception can be guided by the Solidarity view.[5] In this chapter, I outline the four virtues of this view to be developed in the following chapters: (1) respect for bodily diversity, (2) relational authenticity, (3) empowered self-direction, and (4) mutual recognition.

In chapter 2, I address the first two faces of vulnerability, aging and illness, and why we fear them. Often, proponents of genetic enhancements have argued that we can postpone aging and increase our healthspan. I answer that genetic technologies cannot alleviate these two facets of the human condition. Aging and illness disrupt our home life, our personal space, our activities and habits, and our sense of time (Carel 2018; Sarton 2007; cf. de Beauvoir 1972, 698 and Young 2005, 158). I argue that there is an additional reason why we fear aging and illness: because capital accumulation requires exploitable bodies for the labor market, and the market marginalizes those it deems unexploitable.

In this chapter, I introduce Iris Marion Young's five faces of oppression to politicize the injustices and show that genetic enhancements will not alleviate these fears. Drawing upon the Solidarity view, I then develop the virtue of embodied diversity to rectify the potential oppressive structures in our present and future society.

Chapter 3 addresses death as the third face of vulnerability. Part of our desire for an invulnerable body is that we fear facing our own death and the deaths of our loved ones (May 2017). While many proponents of enhancement have suggested the capacity of health as a primary good for enhancement, I argue that there has been a confusion about the powers we attribute to scientific tools. At the source of this confusion is what Marx would call idolatry (Fromm 1961; Gabriel 2015).

Instead, I argue that we should face our own mortality with courage, which is enacted through the second virtue of solidarity: relational authenticity. This is accomplished through the sharing of our life stories with others. In this discussion, I draw on the stories of those who have faced death, which include personal stories and the testimony of those who have made their stories public.

In chapter 4, I return to my journey through infertility as I consider the fourth face of vulnerability, which is misfortune. The fear of failing to achieve aspects of one's life plan can be devastating. In this chapter, I weave together what I have learned from my experience with adoption, and that genetic technologies cannot prevent the face of misfortune from entering our lives. Often, philosophers link the face of misfortune to disability as a way to argue for the elimination of certain traits. In response, I de-link the conception of living with a disability from an unfortunate life by examining its colonialized past.

Then, I critique the culture of motherhood and pregnancy by drawing on the stories from Heather Lanier (2020) and Denise Sherer Jacobson (2007). I argue that much of what we desire in life Marx would consider imaginary appetites, which were created by industry. In our pursuance of these appetites, we commodify human experience. As an alternative, I propose the virtue of empowered self-direction as a way to encounter the face of vulnerability that is misfortune.

In the final chapter, chapter 5, I explore the final face of vulnerability which capitalism exploits: the face of suffering. In this chapter I return to

Kyle's story to address the argument that genetic technologies can relieve this facet of the human condition. Often, philosophers turn to measurements of quality of life (QALY) to determine well-being. Some argue that we have a moral obligation to prevent people with disabilities from being born.[6] Others, such as Peter Singer (2009), argue that those with disabilities should end their lives as soon as possible.

In considering these arguments, I bring into discussion the perspectives of Kyle, Lanier, and Harriet McBryde Johnson (2020) and ask why do their perspectives not count in this discussion? The answer, I think, can be found in two sources of confusion. The first source of confusion concerns the disability paradox.[7] When transhumanists abstract human enhancement from the particular circumstances of social life which give rise to the concrete claims of theories of well-being, they falsely assume that one can distinguish objective claims of well-being from the socially specific claims given for a life well lived.[8] The second source concerns a confusion of pain with suffering. The utilitarian calculus of pain and pleasure quantifies the existential experience of suffering as pain. I argue that this is not only reductive but misses the target at which suffering aims. In both cases of confusion, an epistemic injustice for people with disabilities has occurred.

To rectify the epistemic injustice committed against people with disabilities in this discussion, I propose the virtue of mutual recognition. The Solidarity view expands upon the view of well-being, insofar as my ends are realized through the other as interdependent mutual recognition. The interdependence of our narrative and social identities with our relations with others and our society shape, mold and alter the meaning of how we conceive of our well-being. It is through our solidarity with others, I conclude, that we are able to conceive of the good life.

NOTES

1. The arguments used in genetic enhancement debates concerning disability often discuss contemporary advances in science. When choosing to use reproductive technology, potential parents often desire to have a child that is biologically related. While adopting a child to avoid passing down a particular disease would seem to be a straightforward solution, the decision to use reproductive technology instead demonstrates how strong the desire to have a biological child of one's own is. There are many biological contributors that go into reproductive technology: donation of eggs, sperm or gametes, women who lend their wombs as surrogate mothers, and individuals who supply the germ-lines from which replacement genes are developed. The focus in this book is on testing and engineering the embryo rather than the fetus. The most direct approach would be to select the "best embryo" and then to genetically

modify the problematic genes before conception takes place to avoid an undesirable condition. This procedure, at present, is only theoretical. In practice, it would be extremely complicated: the replacement gene would need to be placed on the right spot on the right chromosome (Wachbroit and Wasserman 2003). In the future, however, this procedure may be achievable for some genes, and at that point it equips parents with the option of eliminating certain diseases or disabilities for their future child as well as possibly enhancing other traits in that child. These alterations would also be passed down in the child's progeny. In sum, genetic engineering technologies raise important questions about whether there should be limits on the choices parents make about the kind of child they have (ibid.).

2. Philosophers disagree whether perfection should be the ideal pursued (Parens 1995, 2005; Mehlman 2009, 2012; see also Annas *et al.* 2002). Perfectionist theories of well-being (e.g., Aristotle) seem to conflate what is good for the individual person and what is morally good as the same. This conflation arises in particular for moral enhancements. The general assumption is that increasing one's cognitive and moral capacities will not only make the individual's life go better, but also will enrich the well-being of society as a whole (e.g., Azevedo 2016; DeGrazia 2014, 2016; Douglas 2010, 2014a, 2014b; Hughes 2014, 2015).

3. There are two ways philosophers discuss genetic engineering in the contemporary debate. First, philosophers often distinguish treatments as therapeutic measures from the enhancement of certain capacities. A therapeutic use of genetic engineering is usually considered a medical treatment. By contrast, a functional enhancement would improve a specific capacity or trait. The common problem facing functional enhancements is the problem of trade-offs. For example, improving one's eyesight may also bring the drawback of being more sensitive to light. Deciding how to navigate these trade-offs consists in weighing potential harms and benefits. The second way that philosophers discuss genetic engineering is by thinking of human enhancement as an ideal. For these discussions, the aim is to enhance one's overall well-being. These philosophers argue that disabilities are an impediment to one's well-being, and as such, should be eliminated if the technology is able to do so. Although many philosophers within the enhancement debate have argued that their aim for human enhancement as an ideal could fit with any of theory of well-being (hedonic, desire satisfaction or objective list), the ideal of human enhancement itself has been criticized for pursuing perfectionism.

4. The genetic enhancement debate, whether for the enhancement of functional capacities or for the ideal of human enhancement, encounters difficulties when considered through the lens of disability. In general, philosophers are divided on the topic of well-being, even more so when addressing the subject of disability. On the one hand, some hold that disability has a negative impact on one's overall well-being, and as such, disabilities should be eliminated from embryos if we have the scientific ability to do so (see Markens 2013). On the other hand, others hold that it is oppressive social and environmental factors that lead to lower well-being. In the debate, disability is sometimes conflated with disease. Within contemporary society, it is difficult to distinguish disability from disease. For example, in the United States, the Americans with Disabilities Act provides legal protections and requirements of reasonable

accommodation. Most diseases are classified as disability to protect individuals from discrimination. Additionally, some health conditions s can cause mobility challenges or can lead to further impairment. Finally, societal accommodations do not provide an environment that is accessible to many forms of human variation. In short, individuals with disabilities are thought to have "atypical functionings" so the focus in medicine should be on prevention or treatment.

5. Many of the philosophers within the enhancement debate adopt a welfarist conception of the good life from John Rawls (1971, 1985). These positions are often hold biased conceptions of the good life and distributive justice for people with disabilities (Wasserman 1998).

6. With respect to the definition of disability, many of the philosophers who want to eliminate disabilities hold to a "normal function" or "species-typical" account of embodiment (Schermer 2008; cf. Hogle 2005; Silvers 1998; Silvers and Stein 2003). This view assumes that physical or mental variation is biologically or statistically based (Amundson 1992, 2000; see also Carlson 2013). The normal function account, however, is widely recognized for not depicting human life accurately. As Buchanan *et al.* summarize, "[a]lthough every impairment of normal functioning constitutes an inability (to perform a normal function), not all impairments result in disabilities. Impairments become disabilities in one sort of social environment but not others" (2000, 287; cf. Silvers and Francis 2010; Corker 2001). In practice, moreover, normal function accounts are not consistent: they forget that certain treatments such as vaccines cause one's bodily functions to elevate above a normal range to produce a defensive response. In this sense, a vaccine would be considered a malfunction or abnormality, while health is considered a functional system. As a result, the position is not only theoretically problematic, but also practically unworkable.

7. The disability paradox results from not taking into account the subjective perspective of those living with a disability. Often, there are two different conceptions of disability at play: the medical model and the social model. Social-model theorists make several claims about embodiment and living with a disability. First, since disability is caused by the social environment, then a society should modify that environment to accommodate and ameliorate the disadvantages for people with disabilities to a degree (Bickenbach 1993). Second, because noted exclusions and disadvantages are caused by social arrangements and practices, a society has a moral obligation not to create or enhance those disadvantages (Wasserman and Asch 2012). Third, a society has a greater obligation to alleviate disadvantages that result in prejudice or stigma, which lead to a greater injustice, than it does to alleviate disadvantages which are neutral. For general reference on the social model, please see Tom Shakespeare (2001). The social model failed to address the concerns of intersectionality, and thus disability scholars such as Robert McRuer (2006) and Carrie Sandahl (2003) developed intersectional models with queer studies and disability as *crip theory* and Nirmala Erevelles (2011) and Alison Kafer (2005, 2013) have brought intersectional issues to bear on feminist concerns. Erevelles has developed a feminist Marxist perspective, which draws on insights from post-colonialism and critical race theory. Kafer, by contrast, has developed her account of embodied subjectivity based on a reconception of disability as a temporal-being-in-the-world (*crip time*) that constructs disability

as desirable and views disabled people as having a future. Ingunn Moser (2006) develops an intersection with disability and class. Most recently, David T. Mitchell and Sharon L. Snyder (2016) have proposed a theory of disability as materiality as an intersectional theory of disabled bodies and their ecological toxic environments. For discussions concerning disability, critical disability theory, and genetic enhancement, consider Hamraie and Fritsch (2019), Nelson, Shew, and Stevens (2019) Carla Rice *et. al* (2017), and Davies and Taylor-Alexander (2019).

8. Julian Savulescu and other transhumanists such as Anders Sandberg and Nick Bostrom, instead, choose to define disability differently by what they call a "welfarist" conception. Their approach views disability as an impediment to well-being by definition: "Any state of a person's biology or psychology [X] which decreases the chance of leading a good life in circumstances C" (2005; see chapter 2 for the 2008 version). According to Savulescu, "[d]isease is a disability to the extent it reduces a person's well-being. Whereas disease is a naturalistic concept, disability is an inherently normative one" (2008, 64). By stipulating a definition of disability as an impediment to well-being, transhumanists argue that any biological or psychological condition (or trait) should be removed from the embryo. What is at play in both the "normal function" and the stipulative "welfarist" accounts is the medical model of disability and it is neither the only nor an incontestable avenue of approach (Parens 2013; Scullion 2010). The medical model views disability and disease from the standpoint of impairment, bodily malfunctioning, or the loss of health. See also Bostrom and Ord (2006), Bostrom and Roache (2008), and Boyle and Savulescu (2001).

Chapter 1

Solidarity and the Five Faces of Vulnerability

INTRODUCTION

Let's talk about the monster under the bed that every parent and future parent fears: the possibility that one's child will be excluded from, or even further *harmed*, by other members of one's society. Disability, and especially serious disabilities, may place one's child in harm's way—or even at face value, a disability might be considered by some a harm in itself.[1] It is this fear that drives many to consider selective reproductive procedures in order to ensure a child of choice. In fact, we believe strongly that the "preventing serious disabilities appears prima facie to be an incontrovertible good" (Kittay 2019, 58). Thus, the expectation is that parents should make every reasonable effort to prevent (physical or psychological) harm coming to their children (or future children).

Genetic technologies, including enhancement technologies, promise to slay this monster. They do so by offering parents a sense of control. Those in favor of genetic technologies such as functional enhancements envision a possible future and see these technologies as a way to rectify our present societal ills. For example, the transhumanist Nick Bostrom has argued that what he calls "Deathist" stories and ideologies are stories and ideologies that "counsel passive acceptance" and "are no longer harmless sources of consolation. They are fatal barriers to urgently needed action" (2005b, 276). Rather than resign ourselves to biological fate, we should instead embrace the possibility to choose a more desirable future. Hence, these technologies promise a source of control: a source of control over the health, talents, and well-being of one's future child. Isn't this what every parent and future parent wants?

This promise, I argue, is false. Let me shine a light on the monster under the bed: I am the situation all parents and future parents fear. My mental health condition is not genetic; it was caused by trauma. And it was not just one event, but rather a series of repeated episodes over my lifetime since childhood. The closest description to my mental health condition would be called Complex Post Traumatic Stress Disorder (C-PTSD), but as my former psychiatrist said, I am "a unique snowball." I think the simplest response to this false promise is the answer Eva Kittay offers: choice is an illusion when we buy into the myth of control in these matters (2019, 59–60).

The myth of control is what supports the dichotomy of chance and choice in the literature on genetic technologies. The myth of control that these new technologies offer, however, does something more: it conceals our fear of vulnerability. This vulnerability shows up in our lives in many forms. Sometimes physical pain or illness remind us of the fragility of our embodiment. At other times, our emotional lives can suffer when relationships come to an end, or if significant projects we've cherished cannot come to fruition. Finally, vulnerability makes itself present when our lives are disrupted or upended by a crisis. In each case, it raises a basic question: how are we to cope with the various faces of vulnerability which show themselves through age, illness, death, misfortune, and suffering and which challenge us to reconceive of our relations to others in our social world (cf. May 2017, 4)?

Control, the transhumanist option, is only one response to what might be called the *five faces of vulnerability*. My aim in this chapter is to establish an alternative perspective on the fear of vulnerability embraced by many enhancement proponents. One important purpose of Marxist praxis is to provide an alternative vision of social relations within society; one which provides a positive conception of the good, inspires mutual respect and recognition, and can motivate action for social change. Many philosophers on either side of the genetic enhancement debate have overlooked "the particular circumstances of social life that give rise to concrete claims of justice" (Young 1990d, 4).

The alternative vision I propose, by contrast, provides a way to reflect on the benefits diversity and social differences in our existing concrete circumstances bestow. Further, it can galvanize us to change the flaws of our societal structures and institutions rather than misguide us into thinking changing society's members will correct societal ills. Changing the bodies of the members of a capitalist society does not alter the structure of capital itself.

In this chapter, I discuss which factors drive the adoption of the myth of control and how the structure of capital needs exploitable bodies from which to accumulate wealth for the few. While philosophers in the enhancement debate have been distracted by the possible trade-offs at the level of the individual, I instead address how capitalism utilizes the fear of vulnerable bodies, which is based on a form of false consciousness and results in a sense of alienation for the laborer. The Five Faces of Vulnerability provide the motivation for the adoption of these technologies. Yet, these technologies, especially genetic enhancement, will only serve as a temporary salve for the five faces which provide the structure for this work. The oppression of bodies which can neither be exploited nor commodified, then, becomes capitalism's solution to its perceived problem. In response, I explain the general conception of the Solidarity view and then sketch the four virtues of this view to be developed in subsequent chapters: (1) respect for bodily diversity, (2) empowered self-direction, (3) relational authenticity, and (4) mutual recognition. Perhaps, though, in order to address the fear of vulnerability it is best to begin with a story.

CONTROLLING VULNERABLE BODIES

In response to proposed abortion legislature in Texas, Rebecca Cokley writes that the "right to decide what happens to our bodies is a fundamental principle in the disability community, and with good reason" (2020, 161). In her essay, Cokley describes the challenges her parents faced conceiving a child. Both of her parents had achondroplasia, which is, according to Cokley, "the most common form of dwarfism" and has a correlation with "babies dying shortly after birth" (ibid.). Before her parents had her, they had experienced "three pregnancies, three baby showers, and three losses" (ibid.). As Cokley notes, these traumatic experiences put strains on their marriage, and as a result, her mother "was unwavering in her support of abortion. A person should have the right to choose" (ibid.).

The debate concerning the use of selective reproductive procedures, especially genetic enhancements, adds another dimension to the discussion concerning bodily choice. In *Learning from My Daughter,* Kittay observes that there is cause for concern with the engineering and enhancement technologies insofar as they aim to eliminate forms of human diversity thought undesirable (Kittay 2019, 67). While the choice of whether or not to become a parent is a personal choice, which requires epistemic humility for others to respect, the choice about using engineering and enhancement technologies offers future parents something different: the choice to control the body of one's future child.

Unfortunately, medical counsel both in the past and present, has had inadequate knowledge or held biased beliefs for making an informed choice about living with a disability. Specifically, it has used the fear of vulnerability to direct the decisions of future parent(s).[2] For example, future parent(s) can have ambiguous feelings about a "fetal abnormality," and uninformed counsel can have an effect on difficult parental decision making, such as selecting to terminate a pregnancy after the prenatal test is positive for Down syndrome (Bijma, van der Heide, and Wildschut 2008).

Licia Carlson cautions that one must be suspicious of those labeled as "experts." According to Carlson, those in the expert role are also gatekeepers of knowledge. In this way, they determine who has the authority to speak about a particular subject and what counts as a knowledge claim (2010, 556; 2009). The assumption is about the future child's quality of life: both the parents and the medical counsel assume that a child with a disability will have a poor quality of life, and the conclusion to terminate the pregnancy is uninformed (Segal 2010). The result, as Melinda Hall says, is a valorization of some bodies over others (ibid. 2017, 66). This valorization, though, is animated by the underlying fear of bodily vulnerability.

This fear of vulnerability arises from an ideal of perfecting the body. For example, in North America, "cultural practices foster demands to control our bodies and to attempt to perfect them, which in turn create rejection, shame, and fear in relation to both failures to control the body or deviations from body ideals" (Wendell 1996, 85). What is implied is the boundary between acceptable bodies and those deemed not acceptable. Bodies which show signs of vulnerability are those most often rejected, feared, or ignored (cf. Wendell 1996, 85).

The ideal of bodily perfection, however, is related to "the economic processes of a consumer society" (Wendell 1996, 86). This ideal generates profits, and similarly the fear of falling short of a desirable body leads to the feelings of shame, self-loathing, and low self-worth. This pursuit of perfection is not just a quest within the media and cultural practices alone, but it is also present within the medical profession, which seeks to control and "cure" the body from various physical and psychological vulnerabilities. As Susan Wendell notes, "the disciplines of normality like those of femininity are not only enforced by others but internalized. For many of us our proximity to the standards of normality is an important aspect of our identity and our sense of social acceptability an aspect of our self-respect" (ibid., 88).

When we pursue perfection, we encounter the fear of being abnormal and shunned, and the fears of pain, illness, suffering and death. Capitalism fosters ignorance of the lived experience of being vulnerable. This ignorance leads to the further marginalization of those with disabilities and illnesses. As Wendell observes, "curiosity about medical diagnosis, physical appearance, and the sexual and other intimate aspects of disabilities is common; interest in the subjective experience is rare" (ibid.6, 91).

What is behind the ideal of bodily perfection is the myth that the body can be controlled (cf. Wendell 1996, 93). By believing this myth, people can psychologically escape the fact that our bodies are vulnerable. This myth of control perpetuates the faulty beliefs that illness, disability, and death can be prevented, or in the very least, postponed (cf. De Grey 2003).

There is, however, an element of truth in this myth: we do have some control over the physical risks in which we engage that could damage our health. The reason that the myth of control is a myth, though, turns on the way it promotes the faulty ideal of invulnerability. That is to say, it supposes that "invulnerability," whether physical or mental, is achievable, or if not, it should be sought after all the same. The myth of controlling the body is part of a larger myth in modern Western science: the myth that nature can be totally controlled (cf. Wendell 1996, 94). These two myths, that nature and the body can be controlled, collide in the debate concerning the use of genetic technologies, especially regarding genetic enhancements.

THE FALSE PROMISE OF GENETIC TECHNOLOGIES

The possibility of controlling nature and the body of one's progeny is tempting. If we think about all of the protection and potential a future parent might choose from, how could we say that their choice to utilize such technologies could be misguided? Often, philosophers, scientists, and those within the medical profession talk about the therapeutic measures, which have the potential to eliminate or alter one's genetic makeup. If you had the possibility to eliminate or at least reduce the chance of developing cancer, contracting HIV, or carrying the gene for Tay-Sachs, wouldn't you take that risk? And moreover, wouldn't you want to protect your child from these chances if you could?

This is precisely the risk that researchers took using CRISPR-Cas9 is the case of Lulu and Nana in China, which were claimed to be the first genetically modified babies with resistance against HIV infection (Khan 2019). It might be argued, however, that Lulu and Nana were enhanced babies. The reasons in favor of this position include the points that Lulu and Nana did not carry a trait for Down syndrome or have a deletion of genetic material from chromosome 7, which is what occurs in Williams syndrome. Rather, Lulu and Nana received what one might call a "booster" in their genetic modification, which made their immune system more resistant to HIV infection. From this perspective, increasing one's healthspan, lifespan, or other genetic capacities is an enhancement.

An initial background sketch might be helpful here. The enhancement of some capacity or trait such as vision, memory, intelligence, health or lifespan is usually referred to as a *functional enhancement* within the literature. *Human enhancement*, by contrast, refers to the enhancement of a human being's life (Veit 2018, 1).[3] Those in favor of enhancement often use both forms of enhancement in discussion, but usually focus on the second sense of enhancement, i.e., human enhancement. This version of enhancement centers on an idealized view of human life and well-being. David Gems (2011), for example, has argued that the enhancement of lifespan (a functional enhancement) should be pursued energetically because decelerated aging would provide many improvements to health despite the transformation of society that it could entail. The additional improvements would likely increase one's overall well-being, and therefore be considered in a discussion on human enhancement.

Those in favor of enhancement include a group of proponents who call themselves transhumanists. Transhumanism views human nature as a "work-in-progress" that could be redesigned in desirable ways. Transhumanists see enhancement technologies as a bridge to a better future, and anticipate that any challenges which may arise, science will be able to correct. In short, the transhumanist vision is to create the opportunity to live much healthier and longer lives, "to enhance our memory and other intellectual faculties, to refine our emotional experiences and increase our subjective sense of well-being, and generally to achieve a greater degree of control over our own lives" (Bostrom 2003, 493–94).[4]

Bostrom (2003) speculates that many more goods are likely to come from pursuing the enhancement endeavor such as the possibility of genetic enhancement leading to better treatment of people with disabilities because there would be a general "demystification" of the genetic contributions to human traits. Moreover, "a decreased incidence of some disabilities could lead to more assistance being available for the remaining affected people to enable them to live full, unrestricted lives through various technological and social supports" (ibid., 498). Further, genetic engineering has great potential to eliminate, or at least alleviate, unnecessary human suffering (ibid., 499). His optimism about the potential of these genetic technologies for some might be infectious: "being free from severe genetic diseases would be a good, as would having a mind that can learn more quickly or having a more robust immune system. Healthier, wittier, happier people may be able to reach new levels culturally" (ibid., 498). Bostrom has wholeheartedly bought into the promise of a possible future these genetic technologies could bring.

At this juncture, I only want to discuss the possibility of therapeutic and functional enhancements because it is these selective technologies which offer the false promise to potential parent(s).[5] For our current therapeutic work on human embryos and clinical trials for AIDS patients, for example, some challenges are arising for treatments. For AIDS patients, these tradeoffs include nuclease off-target effects, which can result in chromosomal rearrangements, i.e., one of the hallmarks of cancer (Koo, Lee and Kim 2015, 476). Though, for cancer patients, the tradeoffs are autoimmune reactions (Jordan 2016). In the case of using gene-editing tools such as CRISPR-Cas9, sickle cell disease (Sanders 2021) and the gene for deafness have been the current targets in human embryos. Specifically, in the case of gene-editing for embryos, the technique has caused a series of problematic side effects: causing cells of human embryos "to discard large chunks of their genetic material" (Wu 2020). Thus, as of now, the transhumanist vision of genetically enhanced post-humans is a long way off.

So, first, the actual therapeutic and enhancement treatments that could be offered to future parents are questionable because there is an assumption that there will be few if any tradeoffs. But there are further questions we should ask which are existential in nature: why do we want to alter the possible futures of our children? And further, what does that desire for alteration—or even perfection—tell us about the values of the society in which we live? Finally, what does it tell us about our own fears of bodily vulnerability?

Critics of enhancement technologies frequently raise concerns about the possible negative side effects for the self-direction of offspring whose parents choose to utilize such technologies (e.g., Habermas 2003).[6] Offspring, however, do not get to decide whether or not to be enhanced. As a result, they appear to be cheated, existentially, out of their future self-direction and lack the ability to create themselves as others do (Habermas 2003; Elliott 2011).

Michael Sandel, for example, holds that our abilities are gifts, and genetic engineering would rob us of the contingency of those gifts. It not only harms the child but also corrupts the practice of parenting: "Whatever its effect on the autonomy of the child, the drive to banish contingency and to master the mystery of birth diminishes the designing parent and corrupts parenting as a social practice governed by norms of unconditional love" (Sandel 2007, 82–83). What he calls "eugenic parenting" is morally problematic because it holds to an aim of mastery and domination rather than appreciating the "gifted character of human powers and achievements" (ibid., 83). This corruption will in turn strip away the "awareness that none of us is wholly responsible for his or her success" which is what protects a meritocratic society from assuming that one's success is what makes one more deserving than others (ibid., 91).

In many ways, I share these critiques of enhancement technologies. The liberalism transhumanists appeal to has no place for the recognition of bodily difference and social identity. The worrisome prospect these genetic technologies raise for one's social identity concerns the valorization of some bodies over others, rather than a respect for bodily difference. If one's society is not equipped to deal with the aim of genetic technologies, then one will only run the risk of exacerbating already existing inequalities, discrimination, and oppression (discussed further in chapter 2). In short, if one does not consider the societal factors when utilizing genetic technologies, one runs the risk of failing to imagine a future child's life as a whole (e.g., Kamm 2005; see chapter 4).

I think, however, that all of these criticisms of enhancement technologies can and should be made without an ableist bias. Too often contemporary discussion of this debate sets up an exhaustive strawman dichotomy between those in favor of enhancement and those against it. The result is that the rejection of one stance logically yet falsely entails the embrace of the converse position. The teeter-tottering of arguments for and against genetic enhancements thus leads to solutions so divorced from daily living (e.g., moral status)[7] they forget the very reason we should be concerned about these technologies in the first place.[8]

My suspicion, then, is that those in favor of enhancement technologies are proponents because they fear the various faces of vulnerability. The worry should not be about whether the technologies would cause damage to our authenticity, talents, or life plans. Rather, the real concern is how capitalism will utilize our fear of vulnerability for the benefit of the few at the expense of the many. Thus, in our desire to prevent physical and psychological harm for our future children, we may instead be placing them in harm's way of greater existential alienation.

ALIENATION AND THE FIVE FACES OF VULNERABILITY

The dilemma facing any future parent considering these technologies is whether to "opt in" or to "opt out." This is not a choice that those who conceive naturally face at this point, but it might be one to worry about in the future.

The Five Faces of Vulnerability

I know what it feels like not "to measure up" and I'm sure you do, too. Not being good enough is something we all fear. It is not actually the quality being measured, but the judgment of others that we fear. Whether it is physical beauty, academic achievement, creativity, musical ability, athletic ability, sociability, etc., there is always at least one area that gets us—the area that causes shame. The pain we feel for our inadequacies, failures, and imperfections drives our decision-making likely more than we'd find comfortable admitting.

Now ask: what would it feel like for your future child? This is the dilemma future parents considering these technologies face. If a parent does not choose to enhance their baby, and instead opts out of having an HIV-resistant baby like Lulu and Nana or something of the like, have they done something wrong? Is it a wrong because their future child will not be able to "keep up" with the enhanced babies of the future?

Like Robert Sparrow and others, I have often wondered if our pursuit of genetic enhancements will lead to an ever-increasing enhancement rat race resulting in further alienation and exploitation (Sparrow 2014a, 2014b). Just as Sparrow, Eric Parens (2005), and Carl Elliott (2011) are suspect, I too am suspicious of the motivations of those in favor of enhancement. Further, I do not think that enhancement technologies will be able to assuage all of society's ills. As Elliot observes:

> A society shapes the identity of its people by reflecting an image back to them. And if that image does not serve as the basis for self-respect and dignity, it can be psychologically damaging. It is a desire to avoid this kind of damage that drives the demand for many enhancement technologies. (2011, 367)

What Elliott has said in the abstract is what parents of children with disabilities encounter in the concrete, namely, the way society judges your children when they are not "normal." As Kittay remarks in *Learning from my Daughter*, the "desire for normalcy" that parents (and future parents) have for both themselves and their children is often balanced against the "fear that a good life cannot be possible for those who fail to fit within the bounds of the normal" (2019, 57). For example, to return to my reflections from the introduction, I witnessed firsthand this bias from the medical profession about the possibility of my eggs having a "defect." The nurse was most likely referring to Down syndrome. Would it have been wrong of me to keep my egg rather than accept the donor's?[9]

There is, however, something else going on in Elliott's remark that we are shaped in the image of our society: if one chooses not to "opt in," then, there is a chance that one's future child will be marginalized by society. Buying into an ideal of bodily perfection and the myth of control results in the marginalization of people with illnesses and disabilities. The chief disadvantage of capitalism is that "many people are unemployed or underemployed and impoverished against their will" (Russell 2019, 14). Buying into such an ideal is what Marx calls a false consciousness.

The fear of a vulnerable child who is excluded from socialization and the labor market might push us to "opt in." Yet, we believe that this outcome is a possibility because our capitalist society has created this outcome. The cultural silence about the elements of vulnerability exacerbates our fears of experiencing them, and thus reinforces the faulty belief that we can control our bodies, and protect ourselves from vulnerability (cf. Wendell 1996, 109).

Vulnerability reveals itself through various faces such as illness, aging, death, misfortune and suffering. These faces are part of our lived experience, but the meaning we give them has been given to us by capitalism. For example, the young, able body forms the site of reference for what is deemed healthy. These five faces of vulnerability are intertwined with societal values, bias and stigma. Through each of them the lived experience we encounter is the experience of "loss."

The lived experience of "loss" suggests an alteration in one's self-direction. What I do not mean by "loss" is a losing to someone else in a competition, nor do I mean "loss" as misplacement or poor judgment. What I mean by "loss" in part concerns losing something or someone of value. Importantly, this sense of value that is lost can be a part of one's sense of self. As a result, it can indicate a loss in one's sense of identity, one's self-trust, and one's self-worth. "Loss" entails a stripping away of a part of one's future or a reinterpretation of one's past. And it is expressed as grief, regret, or sorrow.

Loss has an effect on our expectations. What I believed, desired, assumed or intended, I can no longer do or expect to happen. This can lead to a loss of trust in myself, in others, or in my surroundings. Whether the loss occurs with my body, my personal strivings, or my identity, loss changes my being-in-the-world.

As a result, I may develop a negative self-image and cynical expectations of myself and others. I may feel prejudice and stigma from those around me. Especially, those whom I once held close to my heart may seem quite distant after the experience of loss.

The five faces of vulnerability I am trying to articulate alter our sense of time and space from the lived perspective. They are structured in time, and hence, can alter how we experience them. Impairment can take place along any of the five faces of vulnerability: illness, aging, facing death, misfortune, and suffering. These can occur as singular lived experiences or can overlap. Moreover, the lived experiences of our loved ones encountering these faces may intertwine with our own life story and plans.

In time, the faces of vulnerability alter our point of linear reference. In the case of illness and aging, they alter our relation from the present to our past. In both the lived experiences of illness and aging, we shift from *being able* to *being unable* in the loss of activities we enjoyed and the sense of whom we were. We grieve the loss of what used to be. In contrast, the face of misfortune disrupts our relation from the present to the future. In the distant sense of time that we perceive as part of our life plan, misfortune disrupts, cuts off, or eliminates the aspect of *some day* because that day will no longer come; rather, that life plan *will no longer be*. We grieve the loss of what was yet to come. Third, facing our own death or the death of a loved one intertwines our past and future by shifting our perspective to what I *have not yet done* and *what I won't be able to do.* Finally, suffering concerns the loss experienced with continual pain, rather than momentary pain. Suffering alters our experience of the present insofar as it changes the meaningful sequence of our daily tasks. These tasks have been placed in sequential order of importance to us, and yet, suffering reshapes or restructures these tasks. This quotidian shift reorients our perception of value in our lives to *whether this is possible today.* Through this restructuring of time, the five faces of vulnerability can fracture our subjectivity insofar as they interrupt the *trying to do* of our life plans and life story.

The five faces of vulnerability are also situated in space and as a result, require us to redefine our life spaces and can force us to adopt new roles. The experience of loss as situated in space disrupts our ability to dwell in the places we were once fond of. It causes us to slip or trip up in the familiar foothold of home.

First, the lived experience of illness may cause one to modify one's environment to accommodate one's symptoms (Carel 2018, 14). If one's illness requires more medical treatment, illness may present as a "loss of place" if one is relocated to certain living facilities. Similarly, in the case of aging, the changes in our motility may cause us to modify our environment, or even alter our location to be closer to family and friends. Third, with misfortune, life plans and personal projects are often linked to spaces in our imagination. Misfortune disrupts those spaces revealing the emptiness and loss of hope that once filled them. For example, in the loss of one's loving relationship, we no longer want to inhabit the same places or rooms. Driving past clinics no longer entered causes us to grieve so maybe we choose a different route to avoid those feelings. Fourth, in the case of facing one's death, one experiences a loss of the familiar world; the space we inhabit seems fragile and uncertain rather than firm (ibid.). Our world is now an unsafe place. Finally, suffering reminds us that our personal space was not designed to ease our physical or mental anguish. For example, most rooms in homes and buildings assume the social frame of the able body (Purcell 2014, 201). In this way, we may be or feel incapacitated in our own home.

The five faces are felt as what Maurice Merleau-Ponty calls an "ambivalent present" in his description of the phantom limb in the sense that the absence of what we wished for is made present (2008, 94). It is both a presence which has become absent yet is also simultaneously an absence which is now present much like a dear friend whom we have lost and whose absence is later made present to us by a familiar reminder. The experience of this loss is relational; we become aware of the absence of one we loved when in the presence of others. Certain rooms, certain smells, a laugh, a look . . . all can remind us of what can no longer be.

In the case of illness, symptoms can provide a "capricious interruption" which reminds us of what we can no longer do (Carel 2018, 42). For aging, it may be the realization that there are many activities that we can no longer do alone without the assistance of others. In the case of misfortune, the shift in one's life plan changes our focus and priorities. For death, the acceptance as *felt* reminds us of what we used to hope for or anticipate will be gone; it is the presence of the absence of that hope. Finally, suffering makes present the absence of tranquility.

Capitalism distorts the five faces of vulnerability because it confuses objects with relations; that is, it reifies our relationships with others. Capitalism, thus, gives meaning to the five faces of vulnerability through promoting fear and competition as well as promising a sense of control.

Genetic technologies assume that this vulnerability is located in our bodies, but it is not. Rather, it is located in our lived experience. This is why changing our bodies will not protect us from the five faces of vulnerability. They will only lead to our further alienation and exploitation. Genetic technologies cannot fix these faces; the faces will only shift because they are how we perceive and interpret our lived experience.

Alienation

For Marx, our social organization directs our consciousness in certain directions and blocks us from being aware of certain facts and experiences (Fromm 1961, 21). It is in this way that we have a false consciousness because we are not aware of our true human needs and ideals. As Eric Fromm notes, for Marx, "[o]nly if false consciousness is transformed into true consciousness, that is, only if we are aware of reality, rather than distorting it by rationalizations and fictions, can we also become aware of our real and true human needs" (ibid., 21-22). Is it really a human need to be smarter, have a better memory, or be more athletic? If not, then whose needs are we meeting when we consider using genetic enhancements, especially in the case of our future children?

For Marx, the mode of production of material life gives the conditions for our social, political, and intellectual lives in general. While our human needs are determined by our relationships and sociality, it is the need of production that places our ability to create and form healthy relationships in chains (ibid., 17). From a capitalist's point of view, the body which is desired is the body that is exploitable. As Marta Russell notes:

> Under Marx's labor theory of value, the basis of capitalist accumulation is the concept of surplus labor value . . . If the worker produced the amount of value equivalent only to her wage, there would be nothing leftover for the capitalist and no reason to hire the worker. But because labor power has the capacity to produce more value than its own wages, the worker can be made to work longer than the labor time equivalent of the wage received. (2019, 14)

The distinction between the actual amount of labor needed and the additional labor time the worker works is the difference between necessary labor and surplus value. The capitalist in interested in appropriating surplus value because it is the source of profit. The secret, then, of the expansion of capital is that it has at its disposal a definite quality of those working as unpaid labor. This situation of the laborer thus leads to alienation. Genetic enhancements do not appear to extricate us from the structure of labor and the resulting alienation because these enhancements were created by capitalism.

While the origin of the term "alienation" (*Entfremdung*) comes from Hegel, Marx refigures the concept based on the initial difference between existence and essence.[10] In the *1844 Economic and Philosophical Manuscripts* (1932) and in *Capital* (1887), Marx develops largely the same argument. The origin of alienation is commodity fetishism, which is the belief that inanimate things such as commodities have human powers (or value) and are able to govern the activity of human beings.

Alienation occurs when the labor of the individual is transformed into a power that rules the person as if by a kind of supra-human law. When one becomes alienated, the world in which one lives becomes foreign or estranged because it is out of one's elective control or affirmation. It is in this sense that our labor, i.e., our work, becomes estranged to us.[11]

Here I want to emphasize the moral dimension of alienation by construing it as the set of impediments to the exercise of one's authenticity and self-direction, or the affirmation of the consequences of one's actions. Impediments to self-direction and authenticity may thus be understood to treat a person as a means, rather than as an end, and so objectify that person (Fromm 1961). The incentives that drive capital operation, of course, do just this. In a capitalist society

> every man [*sic*] speculates upon creating a new need in another in order to force him to a new sacrifice, to place him in a new dependence, and to entice him into a new kind of pleasure and thereby into economic ruin. Everyone tries to establish over others an alien power in order to find there the satisfaction of his own egoistic need . . . Man [*sic*] becomes increasingly poor as a man; he has increasing need of money in order to take possession of the hostile being. (Marx 1932, 140–42)

For Marx, the alienated subject is "a mentally and physically dehumanized being . . . the self-conscious and self-acting commodity" (ibid., 111). The only way the alienated subject relates to the outside world is by having it and consuming (using) it: "The less you are, the less you express your life, the more you have, the greater is your alienated life and the greater is the saving of your alienated being" (ibid., 144).

When a person is alienated from oneself, this person does not experience acting as an agent in one's grasp of the world. Instead, the world (i.e., meaning, nature, others, and oneself) is expressed as something foreign and out of control for this individual (Fromm 1961). What one experiences is a world of objects (even if these objects are creations of one's own), rather than a world of persons and cherished items. Stated simply, the experience of alienation is experiencing the world and oneself passively, as subjected to the environment and objects around it. This experience finds canonical expression in the proletarian who does not own the means to make products (capital), and so must sell her labor to whatever market forces dictate, and for a price that will not compensate her worth. In this way, she is alienated from her activities in the world.

The first reason that the promise genetic technologies make proves false is that those in favor of enhancement have forgotten the societal structures and values embedded in our concrete circumstances. The transhumanist aim for human enhancement is an illusion. Many of the proposed thought experiments about the future abilities of genetic engineering might be the stuff of science fiction (Tonkens 2011). When one considers the empirical evidence so far in these genetic technologies, it becomes obvious that we might be overlooking some serious tradeoffs.

Second, and more importantly, the promise these genetic technologies make proves false because the qualities to be fixed or "cured" are products of societal organization. They do not tap into our real human needs. Thus, if we place our trust in technologies that cannot meet our human needs, then we will be misled and disappointed. Moreover, because super-abled bodies are more likely to be exploited, these technologies more likely will worsen our alienation further. The only way to remedy alienation is to develop an approach modeled in solidarity.

SOLIDARITY AS AN ALTERNATIVE VIEW

The term "solidarity" finds its history in Marxist political theory and is often used in social and political debates as an alternative to paradigms of justice regarding the challenges of redistribution and recognition (Ferguson 2009).[12] Many of the more specific facets of this notion are developed later in the book at the appropriate places, where closer attention is warranted. To introduce the notion, then, I provide here only some of its broader characteristics and an introduction to its four virtues.

It might be best to begin with an illustration before listing the characteristics of this view. I would like to consider the three large international studies the World Health Organization (WHO) conducted over the course of twenty-five years for mental health in various countries beginning in the late 1960s (Watters 2010, 137). WHO's research findings showed that patients outside the United States and Europe had significantly lower relapse rates of mental health conditions. In countries such as India, Nigeria and Columbia, patients had longer periods of remission and higher levels of social functioning than patients in the United States, Denmark or Taiwan. In industrialized nations, forty percent of patients with schizophrenia were judged over the life course to be "severely impaired" while only twenty-four percent of patients in "poorer countries" were considered "severely impaired" (ibid.; Purcell 2016a). The result of WHO's findings presented a paradox for the medical community: how could it be that the regions of the world with the most resources to devote to mental health, such as the best technology, cutting edge medicines, and contemporary research, had the most troubled and socially marginalized patients (ibid., 138)?

During the same period, anthropologist Juli McGruder (2004) studied the families of individuals with schizophrenia in Zanzibar for over two decades and found the families used a different framework for understanding mental health. While the Zanzibar population is predominantly Muslim, McGruder observed that the people often make use of Swahili spirit-possession beliefs to explain the actions of someone not conforming to social norms. This violation of social norms could be as mild as a sister lashing out at a brother, but also as extreme as an individual experiencing psychotic delusions. McGruder found that these beliefs served a useful function in mental health treatment: the beliefs provided a variety of socially accepted interventions and ministrations that kept the individual within the family and kinship group. Unlike the "Christian sense" of "casting out demons," McGruder realized that the families would coax the spirits with "food and goods" or "song and dance." These beliefs had, in turn, unexpected benefits for the individual living with schizophrenia: when the illness went into remission, the person could retake his or her responsibilities in the kinship group (McGruder 2004; Watters 2010, chap 3; Purcell 2016a). McGruder's research provided an interesting answer to this tension in WHO's findings: rather than "curing schizophrenia," McGruder determined that the spiritual beliefs of the families in Zanzibar maintained the individual's status within one's social group. Thus, this group cohesion enabled the individual to effectively manage the course of the illness.

McGruder's research illuminates the need to consider an alternative vision: the Solidarity view. This alternative vision includes recognition of our interdependence, recognition of one's positive status within one's social group, and recognition of individual differences. Furthermore, it offers wise guidance about effective efforts for social justice and societal change.[13]

The Solidarity view, thus, builds upon Marx's theory by drawing from philosophers such as Ann Ferguson (2009), Nancy Fraser (2016), Iris Marion

Young (1990a, 2000), Tommie Shelby (2005), and Alison Jaggar (2001, 2014).[14] All have proposed that solidarity includes a mutual co-belonging, a recognition of individual differences, and a sense of shared empowerment within a society. Rather than embrace "justice" as the guide for political philosophy, many of these thinkers embrace "care" as a strategy to provide a better solution to the problems of redistribution and recognition (cf. Kittay 2002b; Baier 1987).

In response to the five faces of vulnerability, the Solidarity view proposes a sense of wonder, trust and cooperation as the foundation for facing vulnerability. This shift in perspective alters the meaning of the aspects of these five faces. The shift in meaning occurs in two respects.

Testimonial Justice

First, it aims for testimonial justice by taking into account the lived experience and perspective of living with a disability (Scuro 2018). Many feminist epistemologists and philosophers of disability have addressed the influence on knowledge that social, historical, and cultural situations have (e.g., Tuana and Sullivan 2006; Stramondo 2011; Wendell 2016; Carlson and Kittay 2010). Knowledge is obtained by *how* we acquire the elements of knowledge, and this acquisition is shaped by bodily differences as well as social conditions. Often other skills are developed as well. For example, consider individuals with Down syndrome and Williams syndrome. Those with Down syndrome may have difficulty thinking abstractly, but they often have stronger skills in the perception and memory of concrete details (Asch 2003, 323). People with Williams syndrome have better social and emotional intelligence; they also often have better musical ability than many of their able-bodied peers.

If we ignore the testimony of people living with disabilities when we consider using genetic technologies such as enhancements, then we are not getting an accurate picture of what to expect. Disability as a category is a "big tent" as Kittay says, and even if one has a specific form of disability, one has no privileged epistemic access into the life of another person with disabilities (2019). To ignore those with the actual lived experience in these discussions is to commit an epistemic injustice (Fricker 2007; Barnes 2014, 2018).[15]

Cripping Time and Space

Second, acting in solidarity together means *cripping* time and space.[16] The five faces of vulnerability are structured according to normative time and able-bodied space. Yet, the way that we as humans experience time and space is from a lived perspective. Each of these can be articulated in two questions. First, whose time? And second, whose spaces?

For disability scholars, normative time is a linear time. Disability disrupts the normative timeline and cultural understandings of the life course, because it is a social, historical, and cultural product (Ljuslinder *et al.* 2020, 35). Normative time focuses on a linear chain of events, which usually begin with birth, progress through reproduction, and then end with death (ibid.; cf. Halberstam 2005).

There are two problems with the normative timeline. First, this linear time structure is normatively problematic insofar as it is linked to heterosexuality, reproduction and the family. Second, it is problematic because it is highly focused on labor and productivity. This suggests that "one should transition from child to adult, find a partner, get married, reproduce, work, eventually transition from adulthood to old age, retire and die" (ibid.). Both heterosexual and ableist norms are embedded into this timeline. To progress in a different order would be considered deviant, wrong or ill (Baril 2016). *Crip time,* by contrast, allows for "time travel" (Ljuslinder *et al.* 2020, 35; Samuels 2017). People "move, think, and speak at a different pace"; it is disability culture politics (ibid., 36; Kuppers 2014).

Disability has the power to disrupt the stages of the life course because it can pull us out of the linear, progressive timeline of the stages and tasks we are expected to complete. The idea of a desire for normalcy as ability which permeates our understanding of time "creates a realm of compulsory able-bodiness" (Ljuslinder *et al.* 2020, 36; McRuer 2006; Kafer 2013). The "pejorative term crip has been reclaimed by the disability community and disability academics as both a site of identification and academic inquiry" (ibid.; cf. McRuer 2006; Yergeau 2018). Crip approaches investigate and are critical of the normalcy of ability within time; in response, crip time redefines time. For example, crip time can be considered a shift in mindset. According to Alison Kafer "rather than bend the bodies and minds to meet the clock, crip time bends the clock to meet disabled bodies and minds" (Kafer 2013, 27).

There are three ways to understand crip time (Ljuslinder *et al.* 2020, 36; Baril 2016). It can be the "extra time" needed to perform a task when compared to ableist time; the lived experience of asking for this "extra time" is viewed as a deviation from what is normal. This additional time is not necessarily a slower pace; it may be due to inaccessible buildings or transportation. The second way that crip time can be understood is as extra wasted time because one is either unproductive or does not live up to the norm. Third, crip time can be understood as an analytical tool. As a tool, it understands "flexible temporalities for different people and not one fixed normal temporality" (ibid.).

Acting in solidarity also means *cripping* space. For example, Jos Boys (2018) has argued that in architectural design, it quickly becomes evident who is considered in space and discourse in contrast to whom gets erased.

Following a cue from Rosemarie Garland-Thomson (2011), Boys expands the concept of "misfitting" to discuss the kinds of bodies excluded or rendered invisible in able-bodied spaces. In response to social and environmental factors designed to privilege the able body, Boys proposes a method of cripping spaces (both disciplinary and actual) through an innovate way of doing and thinking (2018, 55).

There are two problems with able-bodied space. First, the designed space "includes an easily unnoticed slippage *away from* the complexity, variety, and differences of inhabitants' spatial experiences and *toward* the designer's own sensitivity to, and interpretation of, those experiences" (ibid., 56). This bias of design comes from the latent assumption that the designer or architect has an enhanced sensitivity and insight, which is projected onto that space. Yet, we must ask, "what's missing?" And note who is both *there* and *not there*—whoever "they" are—as the "included as excludable" (ibid., 57–58). For example, the privileging of spaces and learning forms becomes apparent in their social construction as barriers to living, learning, and transportation for those whose bodies do not conform (Purcell 2014, 202).

Second, there are different ways of being oriented toward objects and others within space. As Boys notes, to "investigate orientations—in architectural discourses and processes, in everyday social and material practices, and in actual built spaces—opens up for inquiry what is noticed and treated as valid within the discipline of architecture, as well as what is ignored or marginalized" (2018, 59). Built in assumptions of the able body as an approach "orients *away* from the complex and inequitable variations of lived experience" (ibid., 60).

In response, there is a way to crip space. In one way, cripping space is to design and understand space for relational encounters between people, spaces, and objects, whose interactions may be contested and contradictory. Spatial contradictions interfere with the privileging of particular bodies and experiences over others (ibid.).

A second way is to design space by starting from "misfitting" and difference (ibid., 61). This strategy includes incorporating our everyday embodied concerns and experience from the standpoint of disability into our modes of being. This occurs by valuing all modes of embodiment as multiple ways of being-in-the-world rather than treating disabilities as "things" or parts (ibid., 62). This includes the fact that "things reveal themselves as neither distinct entities nor archetypical projections, but rather as modes of relevance invested with meaning and embedded in specific situations and bodily modalities" (ibid.). Boys continues and notes that beginning with "the taken-for-granted opens up the basis of our differential being-in-the-world and can give insights into how things might be done differently" (ibid., 63).

The aim is to eliminate presuppositions of what human bodies must be. When certain bodies are privileged, those bodies feel at home in the world. The bodies excluded, then, feel out of place or without a home. As Boys notes, cripping material space begins with the "care and concern for the multiple ways of being human" (ibid., 66). Further, it challenges us to critically and creatively interact with a diversity of bodies and spaces. Both cripping time and space, then, begin with our concrete lived experience as a starting point rather than an ideal world.

THE VIRTUES OF SOLIDARITY

While some philosophers may object that philosophy should begin in an ideal world, I think that we need first to identify where our current society falls short. If not, we run the risk (and future error) of using these genetic technologies without both acknowledging our current problems and also misunderstanding how thinking about normative principles work. As Elizabeth Anderson notes,

> Unreflective habits guide most of our activity. We are not jarred into critical thinking about our conduct until we confront a problem that stops us from carrying on unreflectively. We recognize the existence of a problem before we have any idea of what would be best or most just . . . Knowledge of the better does not require knowledge of the best. (2010, 3)

Figuring out what is best to do requires an engagement with the current challenges of the global society in which we live as a starting point. This starting point begins with an acknowledgement of the biased systemic structures that operate within the global marketplace of bodily difference. Without considering our concrete circumstances, which includes the experience of living with a disability, I just think decisions will be made in haste. As I mentioned in my story in the introduction, when I both used and considered the selective reproductive treatments along with adoption considerations before me, I realized that the last thing I wanted to do was to make a rushed decision. The advice our adoption caseworker had given us was sound advice: to take our time, to educate ourselves, and to reach out to the agency with any questions or concerns. Similarly, these technologies require us to take time for considered reflection. This is why I think that four virtues are important for living in solidarity with others.

Respect for Bodily Diversity

Respect for bodily diversity is a good place to begin. Scott Woodcock (2009) has argued that we have a *prima facie* obligation to promote diversity, and that, in certain cases, it is morally wrong to reduce the diversity of humans. This is not an obligation to maximize diversity. Rather, the obligation to promote diversity must be balanced against competing obligations. As an instrumental value, diversity gives several benefits to the individual and society.

It is in our best interest to promote the diversity of bodies and their flourishing. Using enhancement technologies to create sameness creates an illusion of symmetrical relations. Iris Young reminds us that symmetrical relations obscure difference:

> People who are different in such social positionings are not so totally other that they can see no similarities and overlaps in their lives, and they often stand in multivalent relations with one another. It makes little sense, however, to describe their similarities and relations as symmetrical, as mirroring one another or reversible." (1997, 346)

Promoting diversity is valuable to a community in search of the truth. As Woodcock argues, diversity is beneficial because it gives epistemic advantage, and because it provides a diversity of perspectives which aids in the social distribution of seeking knowledge (2009, 262–63).

Second, diversity stimulates critical self-reflection in the moral community for members to consider different viewpoints from their own. Bodily difference encourages people to consider their own beliefs and values. Human variation is "instrumentally beneficial with respect to our shared interest in effective rational inquiry" (ibid., 264). It is through "my interaction with others I experience how I am an 'other' for them, and I internalize this objectification to myself through others in the formulation of my own self-conception" (Young 1997, 348). Their perspective of me is different from my own perspective of myself. According to Young, this "relation of self and other, however, is specifically asymmetrical and irreversible, even though it is reciprocal" (ibid.). For moral respect for bodily diversity, then, "each party must recognize that others have irreducible points of view and active interests that respectful action must consider" (ibid.).

Relational Authenticity

Having CPTSD, two qualities I learned to value were self-respect and self-trust, which have never been qualities listed on the enhancement to do list. Yet, I found them essential in order to live authentically with others and to value the gifts I had. The second virtue includes these two qualities, and I call it relational authenticity.

First, relational authenticity enables us to form a point of view that is authentically our own, which is the foundation for self-respect (cf. Mackenzie 2014, 2015). The self-respect gained is fostered through an individual's social relationships and personal history. Often, when we encounter others, our relationships are asymmetrical and interdependent. We learn self-respect by listening to the stories that others tell. This exchange between persons in asymmetrical relationships requires a recognition of our interdependent relations with others. It involves recognizing that we are not alone.

Second, relational authenticity involves building self-trust through telling one's story. Learning to respect oneself and trust oneself, especially when this trust has been lost and must be regained, is developed with others. Barbara Arneil develops this point, arguing that interdependency must redefine dependency as a "constellation of support" and must provide a "gradient scale in which we are all in various ways and to different degrees both dependent on others and independent, depending on the particular stage we are at in the life cycle as well as the degree to which the world is structured to respond to some variations better than others" (2009, 234). Those who provide a constellation of support for us help us build self-trust so that we can tell our own story.

Empowered Self-Direction

Since the self is a product of social relations, one may find a sense of meaning that is shared with other members of one's group (Purcell 2019). In her essay "City of Difference" (1990b), Young's ideal of the city life emphasizes the concept of empowerment, which she defines as the agent's participation in decision-making through exercising one's voice and vote. According to Young, agents "who are empowered with a voice to discuss ends and means of collective life, and who have institutionalized means of participating in those decisions, whether directly or through representatives, open together onto a set of publics where none has autonomy" (ibid., 251). Extending this idea, I argue that we should aim to develop empowered self-direction in ourselves and in others.

At the heart of empowered self-direction is a sense of wonder. A sense of wonder includes a respectful stance of openness to the needs, interests, perceptions, or values of others (Young 1997, 358). It includes a sense of mystery at one's self and another: it is "being able to see one's own position, assumptions, perspective as strange, because it has been put in relation to others" (ibid.). Wonder is what provides the basis for our possible life plans and choices.

Further, it aids in the understanding of our consequences for acting from our experiences and interests by helping us develop and exercise agentic skills. Our empowered self-direction is interlocked in the narrative histories and interests of others (ibid., 359). When we consider ourselves and others with wonder, we open ourselves up to consider different styles and approaches to living, which can guide us in making decisions about using genetic technologies.

Mutual Recognition

Respect for bodily difference, relational authenticity and empowered self-direction depend on the fourth virtue to enact solidarity with others: the virtue of mutual recognition. Often, in debates concerning genetic enhancements, philosophers discuss the possible distributions and allocations of enhanced capacities. The issue with using liberalism to frame the discussion of these genetic technologies is that these debates lead to the misrecognition and misframing of what it means to live with a disability. In *Scales of Justice,* Nancy Fraser observes that,

> representation is not only a matter of ensuring equal political communities; in addition, it requires reframing disputes about justice that cannot be properly contained within established polities. In contesting misframing, therefore, transnational feminism is reconfiguring gender justice as a three-dimensional problem, in which redistribution, recognition, and representation must be integrated in a balanced way. (2016, 114)

Fraser's emphasis on representation requiring mutual recognition is important for the discussions concerning genetic technologies. If people with disabilities are excluded from the conversation, then we are excluding a community of knowers who have situated knowledge concerning the concrete circumstances of which as a society we may be ignorant.

Social identity cannot be adequately measured by individual advantage within the redistribution framework alone. People with disabilities are often stigmatized (Wendell 1989). We must remember the ways in which "social structures shape how people experience the possibilities of forming a family" (Shanley and Asch 2009, 852). Social difference is not identity; rather, social difference is relational in nature (Young 2000, 89).

In the next four chapters, I develop each of these virtues in response to the faces of vulnerability. And in each I show why the debate concerning these genetic technologies is misguided and cannot be separated from politics. An alternative view grounded in the concrete lived experience of diverse bodies and perspectives provides better guidance for our future.

CONCLUDING THOUGHTS

The transhumanist vision does not purify politics but rather avoids politics. In a sense, it is wildly utopian. To bring it into being would require first a dismantling and complete overhaul of the welfare capitalist society. Science will not change the society we have built; it will only add another dimension to it. A model to transform society must begin from the concrete circumstances of our daily lives, which are given to us by material structures in a particular time in history.

The changes in genetic technology are now a fact (Robert and Baylis 2003; Buchanan 2011a). Even if their development were curtailed in one country, there is no guarantee that the technologies would not be developed and used in other countries. Most likely these technologies will be absorbed by the global marketplace which extracts and exploits bodies for financial gain. Without considering the concrete, social and nonideal features of our world, the pursuance of human enhancement as an ideal is nonsense. The Solidarity view offers the only one which presently attempts this task.

With regard to the use of genetic technologies within society, the aim must be to increase societal well-being from the "outside in" rather than the "inside out." Without changing present societal injustices such as oppression and discrimination, genetic enhancement will only exacerbate those societal inequalities. Those in the enhancement debate are right to speculate that enhancement technologies could greater stratify the systemic forces that operate invisibly within society (Buchanan 2011b; Bostrom and Savulescu 2008). Yet, altering the health of future humans will not necessarily lead to greater life expectancy if certain material and psychosocial challenges have not also been remedied. An account of social justice that ignores interdependence, solidarity, and recognition is thus incomplete.

In this next chapter, I address how the fear that genetic enhancements will drive a deeper wedge between the enhanced and the unenhanced in their competition for material and social goods has forgotten that systemic social features structure our relations to one another. Disability is not a property individuals have but a relation between individuals and the environment. The concern is not that people gain respect from others but rather that people respect one another and respect their bodily differences. The Solidarity view recognizes the fact that justice within a future enhanced society must rectify discrimination and oppression. By understanding that justice includes not only equal opportunity but also equal respect, a plan of action and policy can emerge regarding genetic technologies for our future global society.

NOTES

1. Elizabeth Barnes (2009a, 2009b, 2014, 2018) has noted that to cause a disability is a harm, but to have a disability is not a harm; instead, to have a disability is a mere difference-maker. In general, I have to agree with her on this stance. I disagree, though, with defining disability solely according to the social model. I generally use the term *disabled* like Marta Russell (2019) to indicate a dimension of capitalism whereas I use the phrase *people with disabilities* to refer to social identities.

2. Medical historians and sociologists have noted that "health problems" are not objective. Rather, they are often shaped by cultural beliefs and social values (Carel 2007, Savulescu 2009a). For a general history of genetic counseling on this topic, see Aalfs, , Smets, and Leschot 2007.

3. Theories of well-being generally fall into three camps. The first type of theory is (evaluative) hedonism, which holds that humans pursue what they think will give them the greatest balance of pleasure over pain in life. For example, Jeremy Bentham's theory of utilitarianism, which measures the impact of the pleasant and painful experiences throughout life by their duration and their intensity. Some philosophers such as John S. Mill (1861, 1869) include the additional property of quality in order to draw the distinction between higher pleasures and lower pleasures. In contrast to looking at pleasure and pain to flesh out well-being, some philosophers have proposed desire theories. In short, our well-being is determined by the satisfaction of our desires or preferences. Comprehensive desire theory is usually what philosophers have in mind. For comprehensive desire theory, what matters to an individual's well-being is the overall level of desire-satisfaction in one's life as a whole. Basically, the more desire-fulfillment in life is better. While hedonist and desire theories appear similar, they actually differ formally. Both agree that pleasurable experiences make a good life for individuals, but hedonist theories hold that pleasantness is the good-maker while desire theories argue that desire-satisfaction is the good-maker. The issue, then, facing desire-satisfaction is that it forgets about the pleasure experienced during the process. Is the only enjoyable factor of painting a portrait that one finishes? We seem to think that the process is good in itself as well. The third type of theory of well-being is the objective list theory. Objective list theories give lists of items that constitute well-being such as knowledge, friendship or play. The general source of contention for these theories concerns what should go on the list. Objective list theories, however, tend to be elitist and exclusionary, especially viewed from the lens of disability. Martha Nussbaum (2006, 2009) has tried to correct her objective list account of capabilities to include people with disabilities. For a general reference to welfarist theories of well-being, see Christopher Woodard (2013).

4. For example, John Harris (2011) and Eric Juengst (2017), among others, have discussed the ethics of genetic engineering from a broadly transhumanist perspective. See also Chan and Harris (2007, 2011) and Eric Juengst (1997).

5. Functional enhancements have proven to be a double-edge sword in laboratory studies. One example is the enhancement of memory. Memory enhancement using genetic technologies has been utilized in experiments on mice and rats. As "normal" mice and rats mature, as with other mammals, "synthesis of the NR2B subunit of the

NMDA receptor is gradually replaced with synthesis of the NR2A subunit" (Bostrom and Sandberg 2009, 312). Researchers Joe Tsien and co-workers altered the mice to increase the production of the NR2B subunits, which improved the memory acquisition and retention of the "enhanced" mice (Falls, Miserendino et al. 1992). While this alteration seemed to be a benefit, this modification included an unforeseen consequence: this modification also included increased sensitivity to certain forms of pain (ibid.; Wei, Wang et al. 2001). The trade-off for the functional enhancement was not insignificant: although I might have the benefit of an enhanced memory to aid me in acquiring certain skills and knowledge, I also receive the harm of being more sensitive to certain pains. This trade off may not prove to be beneficial to my overall well-being. More studies were conducted that used different methods to enhance memory in mice. In these two studies, researchers either increased the amounts of brain growth factors (Routtenberg, Cantallops et al. 2000) or increased the signal transduction protein adenylyl cyclase (Wang, Ferguson et al. 2004) to enhance memory. These studies had mixed results again. In the cyclase mice, while recognition memory improved, there was no improvement in context or cue learning. Further, unlearning, for areas such as fear conditioning, took longer for these modified mice in both studies. For humans, this could again prove troubling if it is the case that it would take longer to recover from certain traumas or phobias.

6. Stated simply, they worry that human enhancement will strip away the value of authenticity (Sandel 2007; Bublitz and Reinhard 2009). An agent's authenticity is expressed in her ability to act, reflect, and choose based on factors that are her own (cf. Guignon 2004). Human enhancement seems to create short-cuts for our authentic achievements such as the qualities that people cherish about themselves: endurance, determination, growth, faith, even luck. If that is right, then, our victories will become hollow (Lipsman and Glannon 2013). The loss of authentic achievements not only hurts the character of the user (Sandel 2007), but also alienates the agent from herself and those around her (Agar 2014b; Parens 2005). The breaking down of authenticity, then, damages the bonds of our solidarity and connection with others (Sparrow 2014a, 2014b).

7. An argument often raised to prohibit the use of genetic enhancement technologies turns on the possibility of a genetically stratified society between persons (unenhanced) and post-persons (enhanced). Instead of focusing on the exacerbation of already present distributive inequalities of resources, opportunities, and welfare (as Transhumanists generally do), the worry is that genetic enhancement technologies might also lead to a profound inequality with regard to moral status. Some philosophers have argued in response that a difference in moral status will not occur provided that we choose the right account of moral standing. They argue instead that what might emerge is a conflict of rights. Specifically, philosophers are divided on the topic of moral status for genetic enhancement. On the bioconservative side, Leon Kass (2001, 2003), Robert Sparrow (2014a), and Francis Fukuyama (2002) worry about enhancement producing beings with a higher moral status in terms of human nature. Enhancements may create beings who are not human beings, but who are superior to humans in ways that are or might be thought to be sufficient for having a higher moral status. Jeff McMahan (2009), however, thinks the possibility of the

emergence of beings with a higher moral status and its moral implications can be framed without recourse to the concept of human nature. Philosophers in support of enhancement, such as Nicholas Agar (2010, 2014b), are concerned about what Agar calls "radical enhancement" unless the post-persons would be morally enhanced. The fear is that post-persons could annihilate persons. Persson and Savulescu (2008, 2015; Savulescu 2009b) have argued that without moral enhancement post-persons could commit greater evils. Vojin Rakić (2015), disagrees with Agar, and argues that morally enhanced post-persons would only choose to annihilate persons if they deemed it a moral necessity. Johnston et al. (2018), by contrast, question which moral traits should be enhanced in the first place. For an overview of this debate, see (Gordijn and Chadwick 2008) and (Robert and Baylis 2003). Thomas Douglas (2013) and David DeGrazia (2008, 2012) do not think that Buchanan's argument for a threshold account works. DeGrazia (2012), for example, argues for an interests-based account rather than a respect-based account for moral status.

8. People with cognitive differences are usually discussed in philosophical debates on moral status and moral considerability. Some philosophers, such as Singer (2009) and McMahan (1996, 2002), have argued that since we give people with severe cognitive impairments moral standing, then we should also grant moral standing to animals. Not to do so would be "speciesist." Philosophers have responded to these arguments in several defenses. For example, Matthew Liao (2010) argues for the genetic basis of moral status for moral agency against attacks for speciesism. Anita Silvers (1995, 2012), by contrast, has argued that we should just get rid of the idea of moral status because it does not settle debates about our moral obligations to others.

9. For example, Heather Lanier discusses how Richard Dawkins tweeted that it was "immoral for pregnant women to knowingly carry a child with Down syndrome to term because such a choice would decrease happiness and increase suffering. Abort it and try again he wrote to a mom" (2020, 260–61).

10. According to Fromm, the concept is based on the difference between existence and essence; it is when one's existence is alienated from one's essence, when one is not what she potentially should be.

11. Marx's position is historically materialist, and at least one prominent way that view has been developed holds that human modes of interaction are historically conditioned and determined. G.A. Cohen (2000) at one point understood Marx in that way. This interpretation, however, is not only implausible, but also it strips alienation of moral significance, so I will draw from Eric Fomm's (1961) interpretation of Marx primarily.

12. Marx's use of Solidarity differs from Sandel, who writes "If the genetic revolution erodes our appreciation for the gifted character of human powers and achievements, it will transform three key features of our moral landscape—humility, responsibility and solidarity" (2007, 86). Here, I follow Young's conception of differentiated solidarity, which recognizes bodily difference.

13. McGruder's example has been adapted from Purcell (2016a).

14. I have in mind some of the features from Tommie Shelby's account of solidarity (2005, 69–70) such as shared values, goals and mutual trust, but for group

identification I have in mind intersectional social group identities and allyship. Specifically, my focus is primarily on the intersection of class and disability.

15. David Coady (2012) draws a distinction between two kinds of epistemic injustice. The first kind of injustice occurs when an individual's right to know is violated (2012, 105). The second kind refers to Miranka Fricker's development of epistemic injustice. I will primarily drawing upon Fricker's sense of epistemic injustice in this book.

16. Melanie Yergeau (2018) has pointed out that the term crip is both valued and contentious within disability scholarship: "crip is to disability what queer is to LGBT; while each calls upon and is informed by the other, cripness and queerness cannot be reduced to stasis, and both resist the respectability politics that work to make marginalized people prosocial or governable . . . at times remains a point of contention in neurodivergent spaces" (84–85). To be clear, *crip* is not limited to or the same as the term "disability" because it is not a "polar, linear, or diagrammable construct" (ibid.). Similarly, Yergeau argues that the term *neuroqueer* "emerges as an analogue and expansion of *crip*" for further development and activism. I have crip theory and the neuroqueer movement in mind for breaking down the prejudicial barriers in normative time and space.

Chapter 2

Facing Aging and Illness

Oppression and the Exploitable Body

INTRODUCTION

"I am thinking about trying IVF." I talked quietly at a Halloween party with two friends who had just had children, and who are both professors. One of my friends had gone through the same treatments my spouse and I had already tried. She later gave birth to beautiful twins. She immediately said, "Oh no, I couldn't do it." My other friend had become pregnant without assisted reproductive technologies. She became pensive, and said, "you know, it might be the case that more than one embryo takes. You'll have to make a choice at that point. Abort one or not. Before you do it, you might want to figure out how you will decide."

My friend was right. How would I make my decision? What criteria would I use (if any)? Would the criterion of "healthy" be a factor? And if so, what does that even mean?

Many people in the United States would not consider altering an embryo to have "better health" to be controversial. Should we not use the best science and technologies at hand to better our lives and the lives of our future children? Yet, there is a general ethical divide among philosophers who argue that we should eliminate disabilities in human embryos. If it is possible to reduce illness and suffering, by engineering human embryos to be HIV-resistant, then some argue that it is in the interest of public health to do so. Other philosophers argue that such an intervention should not be permitted because it may lead down a slippery slope[1] to the practice of eliminating or preventing all disabilities or all socially perceived negative traits. This aim, they argue, constitutes a form of positive eugenics, and so is troubling ethically for the same reasons eugenics is troubling as one of history's most infamous cases (cf. Juengst 2017).[2]

In the debates concerning genetic technologies, though, we find little engagement with capitalism and disability even though these topics are part of the underlying structures for the development and use of these technologies. Contemporary emancipatory social movements have developed a habit of suspicion; there is a need to analyze and evaluate the social structures behind these technologies. If we do not ask who will benefits, then how can we be sure that what we undergo—and further, what our future children might undergo—truly enhances our well-being and makes our lives go better?

I think that this habit of suspicion is relevant for the debate concerning genetic modification and enhancements. A major political project for disability activists is to identify the societal factors that disable and oppress those whose bodies do not fit the "norm." This is not just a critical engagement with negative attitudes and policies, but also a critical engagement with the ways in which the accumulation of capital demands bodies for labor from which it can extract the most profit. On the one hand, the health and fitness of professional athletes have allowed for capitalist owners to extract immense profits. Similarly, when the costs of those with health needs rise too high, eugenic programs, or simply adjusting the qualifying rate, dictate who receives benefits within a society. On the other hand, if the disabled body can produce a higher profit—as in the case of Assisted Living homes—then the logic of capital dictates that those bodies are valuable as well. The value of one's body within society is determined by how much it can be exploited or for which it can be auctioned. This is why surrogate motherhood is expensive. Wombs are costly. And children even more so.

In the previous chapter, I proposed an alternative to the liberal conception of the good—the Solidarity view. A conception of the good for society should provide a shift in mindset to view vulnerability as a good of the human condition rather than as an aspect to be feared. It should also address the constraints of oppression and domination. These two constraints operate not only along the logic of distribution, but more importantly operate according to the logic of capital, which affects the division of labor and culture.

In this chapter, I offer some explication of the role that oppression and capital accumulation do play and could further play in the debate on genetic technologies. While philosophers who have endorsed genetic enhancements have tried to think of societal policies to offset the imbalance between the haves and the have-nots, they have yet to consider the role that injustice within institutions and at the systemic level could play. In this chapter I consider this dimension within the enhancement debate through the lens of Iris Marion Young's five faces of oppression. I then develop the virtue of embodied diversity from the Solidarity view as a way to rectify the potential oppressive structures in our present and future society. To begin, though, I think it is best to start with the faces of vulnerability operating behind the scenes in the labor market: the faces of aging and illness.

AGING AND ILLNESS AS FACES OF VULNERABILITY

We don't fear disease; instead, we fear illness. When we think about the actual lived experience of having a disease, we think about how disease as *illness* affects the existential and social dimensions of our daily living. Unlike illness, a disease is used to denote a physiological dysfunction (Carel 2018, 1). For example, it may be the fact that we have a disease, but do not experience any symptoms. In this way, its presence is unknown to us in our daily living. Only when disease starts to disrupt our daily activities do we experience illness. We fear illness because of what we anticipate losing. What we fear losing is a part of who we are.

Likewise, we don't fear age; rather, we fear aging. We fear a disconnect between wanting to do something and our body not being able to do it anymore. Whether it is running a marathon, lifting heavy groceries, or being able to drive our car, aging reminds us of what we can no longer have. This kind of loss happens when we grow out of childhood insofar as playing at the park becomes no longer "fun." It happens again in the transition from adolescence to adulthood, when we have to make a decision about whether to have a future career path, and further, if and when to start a family. And it happens again in midlife when we realize our own insignificance because we are both too old and too young to be of the general public's concern. Yet, while these transitions are to be expected, they are not the transitions we fear. The final transition, the one from midlife to "old age" is the one that terrifies us because we fear our dependency on others. We fear losing who we used to be.

The faces of illness and aging are feared because they are attuned to the past in normative time. The shift in time occurs when *being able to* becomes *being unable to* (ibid., 9). This shift occurs in the recognition that we cannot return to our previous way of life.

Both illness and aging cause a doubling effect in a mistrust of our embodiment. First, we doubt our bodies insofar as we experience the world in which we are "unable to do" the activities we used to enjoy. Physically, illness and aging can cause an existential transformation of our lives in an embodied and situated way by breaking down what is meaningful to us. The possibility of our dependency on others may cause us to deny, flee, or resist thinking about our vulnerability (ibid., 5). In this way, we are no longer able to do what we once did. We fear that aging and illness will impair us from doing the "what" of what we used to do.

And second, this loss of continuity and loss of faith in our bodies alters the structure of how we experience the world itself. In normative time, it causes us to mistrust our bodies because we are "unable to be" ourselves (ibid., 83). Being unable to be is coming to think of our existence as being more dependent on others as well as losing our autonomy and freedom (ibid.). May Sarton, for example, writes in her journals that "[i]n the course of aging, all bodies begin to change—slowing down, losing range of motion, and incurring unfamiliar systemic conditions—which may be felt, in a youth-oriented society, as shame-inducing diminishment" (2007, 183). Similarly, for illness, as Havi Carel has observed, the concept of *being unable to* is intimately linked to an ability to be. *Being unable to* becomes a way of existence and reminds us that our *ability to be* is a "fragile, transient gift" (2018, 84). We fear that aging and illness will impair us from being the "who" of who we used to be.

Both illness and aging can lead to an inability to care for ourselves, which can strip us of autonomy and bring on a sense of helplessness and despair (ibid., 78). We no longer have freedom, openness, and the power of self-determination. As Carel notes, "our plans and aims connect present actions to a future view of ourselves" (ibid.). Illness and aging can prevent or constrain our agency in the pursuit of these plans, actions, and projects (ibid., 80). Our future trajectory can become quite different than the one for which we had hoped. In response to old projects being abandoned, we then must create new projects which must be thought out "in light of the limitations" imposed by illness and aging (ibid., 82). Our possibilities are "shaped and restricted to an extent by aspects of *thrownness*" (ibid.). Thus, in the sense of altering who we are able to be and what we are able to do in normative time, aging and illness can fracture our sense of identity.

We also fear the faces of illness and aging because they reveal themselves to us through ableist space; they reveal what the construction of youth and health have long concealed. Aging and illness change how we experience the world and how we inhabit it. In the case of illness, the ill person must now modify her environment around her illness (ibid., 14). Similarly, in the case of aging, the modification of our personal space may need to be a different location or it may mean relying on others to maintain our home (cf. Young 2005). For example, Carel notes that phenomenological concepts such as "being in the world, authenticity, anxiety, uncanniness and the body's lived experience" have been used to describe living with illness in daily life (2018, 37). Illness "transforms one's being in the world, including one's relationship to the environment, social and temporal structures, and one's identity" (ibid.). Thus, illness and aging do not just change our physical world; they change our social world as well.

Carel first received her diagnosis with "a progressive, untreatable, and ultimately fatal lung condition" at age thirty-five (Haliburton 2014, 231). She describes her encounters with medical staff as "alienating, isolating, and dehumanizing" because they focused on her disease rather than on her, the person or subject, experiencing the illness (ibid.). Serious illness, for example, can leave a person stranded, "without a map and a destination" (ibid., 239). Even the world that was once familiar to us is now lost (Carel 2018, 43).[3]

May Sarton describes the lived experience of aging in her journals. Like Carel, she reflects on the particular challenges for a woman who lives alone and has lost both capability and social value (2007, 185). Here, she addresses her "steady accumulations of losses and disappointments": for example, "friends recently dead," encounters with "demeaning physicians," and the many things she "can no longer do alone, or even at all" (ibid., 185–86). For Sarton, she is quite aware of the "youth-, male-, and health-oriented" culture of the United States, and how "growing old as a woman" can leave one feeling "miserable" and "constantly frustrated" (ibid., 186).

As the personal stories of Carel and Sarton reveal, aging and illness do not just alter our perception of time and space; they also alter our perception of social relations (Carel 2018, 71). At the social level, we experience changes in health care treatment and social attitudes toward illness, aging, and disability (ibid., 17). These experiences can lead to feelings of uncanniness, bodily alienation, and bodily doubt (ibid., 18). According to Carel: "the ill person might be unable to participate and reciprocate in events such as inviting people to dinner if cooking is difficult, there may be awkwardness around the subject of illness or disability . . ." (ibid., 71). There are practical problems such as being able to go for a walk or go dancing (ibid., 76). In short, there can be severe damage to our social world. Spontaneity is gone (ibid., 77).

The faces of illness and aging that we fear, then, alter our lived experience in time, in space, and in relation to others. If there were an opportunity to eliminate or reduce the losses associated with illness and aging, should we not elect to capitalize on this opportunity? The opportunity that proponents of enhancements have in mind include enhancing our lifespan and healthspan. The questions at stake for the use of genetic technologies such as enhancements is the following: will altering our bodies, then, prevent or at least postpone the lived experience of illness and aging? Will they actually alleviate our fears of vulnerability? In this next section, I review the hope that genetic technologies can alleviate these fears and the effect of the labor market on illness, aging, and disability.

ENHANCING CAPACITIES FOR THE LABOR MARKET

The faces of aging and illness are aspects of our lived experience. With them, we fear how the change in our bodies will change aspects of our lived experience. Proponents of genetic enhancements hope that by increasing our healthspan and lifespan we can recover control over our bodies to continue doing the activities we enjoy. They hope that enhancements will help us retain the activities we lose by getting ill and growing old. In the case of aging, specifically, enhancements give us the chance to postpone that final transition until we are ready. Yet, is this really what we want?

Enhancing Healthspan and Lifespan

Transhumanists answer this question in the affirmative.[4] Bostrom, for example, lobbies for enhancement options such as "the radical extension of the human health-span, eradication of disease, elimination of unnecessary suffering, and augmentation of human intellectual, physical and emotional capacities" (2003, 493). As he writes,

> Transhumanism does not require us to say that we should favor post-human beings over human beings, but that the right way of favoring human beings is by enabling us to realize our ideals better and that some of our ideals may well be located outside the space of modes of being that are accessible to us with our current biological constitution. (2003, 495)

The desired end, according to these philosophers, then is an increase in well-being and opportunities for everyone. Increased lifespan and healthspan would be part of this increase in well-being.[5] To conquer the face of illness, then, the transhumanists are proposing a new way of being, i.e., the "post-human."

Aubrey De Grey (2003) seems to have a similar proposal in mind for the face of aging, by arguing for a "cure." Both Bostrom and De Grey think that enhancements will be able to eradicate disease and the biological aspect of aging "as a progressive deterioration of an organism over time, wherein the risk of mortality increases exponentially with age in the post-reproductive years" (West *et al.* 2019, 867). De Grey writes, "[c]learly aging is harder to combat than most early-onset diseases—otherwise we would have cured it already—but the suggestion that it is qualitatively harder lacks the slightest basis" (2003, 928). If we compare an "old healthy person" with a "young sick one" we quickly realize that the younger one will live longer because of youthfulness (ibid., 929). In his comparison between the old and young person, De Grey seems to have in mind that enhancements have the potential to resist how aging and illness shift our state of being. Thus, transhumanists think enhancements will overcome how illness and aging lead to our *inability to be.*

In response to the enthusiasm of the transhumanists, some philosophers, however, such as Leon Kass (2001) and Maxwell Mehlman (2012), have doubts that human enhancement technologies will be able to "conquer" aging and even defeat mortality itself. Kass has challenged the enthusiasm of transhumanists:

> How many years are reasonably few? Let us start with ten. Which of us would find unreasonable or unwelcome the addition of ten healthy and vigorous years to his or her life, years like those between ages thirty and forty? We could learn more, earn more, see more, do more. Maybe we should ask for five years on top of that? Or ten? Why not fifteen, or twenty, or more? (2001, 19)

Kass observes that even if we cannot agree on a reasonable number of added years, some enhancement enthusiasts might turn to a reasonable principle to guide their decision. Similarly, Nicholas Agar (2010) has brought up the emotion of fear as an example of how enhancements may not give us the desired ends we hope for: "I suspect that the fear of death may completely dominate the lives of negligibly senescent people. It will do so to such an extent that it will prevent them from enjoying many of the activities that make our lives pleasurable and meaningful" (2010, 114). These philosophers, then, in contrast to the transhumanists, take issue with the other lived experience of aging and illness: the *inability to do.*

Yet, if we consider the possibility of enhancing traits for a better lifespan or healthspan, we are thinking about the possibility of traits that we might possess. In this way, we are thinking neither of doing nor being; rather, we are thinking of having. To illustrate this confusion, when we think of the enhancement of capacities or traits in the concrete, then we realize that we are thinking about specific enhancements that science could someday alter. Enhancements may enable us to have or possess particular capacities such as a stronger immune system or an addition of potential years to our lifespan, but they will not necessarily enable us to exercise (do) those capacities in order to enhance our sense of identity and wholeness (be). When Bostrom and De Grey discuss the future, they are envisioning humans (or post-humans) who possess (have) enhanced healthspans and lifespans. Yet, there is no guarantee that the enhancement of either possessed capacity will lead to increased well-being overall.

This idea of confusing possession with creating or doing finds its home in Marx's philosophy of alienation. For Marx, work is the active creation of a new world, whether the activity is manual, artistic, or intellectual. Within a capitalist society, humans become estranged from their creative powers. The objects they make as workers become alien to them, and as commodities rule over them. These objects become independent of the powers of the producer (Marx 1932, 42). Whether or not humans have an enhanced capacity, alienation prevents the proper exercise of one's creative powers within a capitalist society.

The philosopher Iris Young (2000) develops Marx's distinction between having and doing with regard to social justice. The distributive paradigm of justice focuses on the possession of, or the having of, material goods and social positions. Beyond material goods, there are also social goods such as self-respect, opportunity, power and honor. Young, however, notes that a confusion occurs here: the logic of distribution treats nonmaterial goods as identifiable things. This conflation of material and social goods reifies the social goods and the individual.

This distinction between doing and having is likewise important for the use of genetic technologies for enhancement purposes and increased well-being. Well-being results not only from having capacities but also from exercising these capacities (doing), that is, the doing of the joyful or meaningful activities; similarly, I consider these activities as an expression of myself and my interests (being). Merely having the capacities, entails neither that the individual will use that capacity nor will I identify with it. It will not alleviate the lived experience of aging and illness, which are tied to being *unable to do* and *to be,* not *unable to have.* To conceive of the mere possession of enhanced capacities then, as a state of being and doing, is to commit a category error.

Second, because aging and illness are aspects of lived experience, they are not puzzles to be solved like medical phenomena such as specific diseases. As aspects of lived experience, they exist in relation to the perspective of those experiencing them. For those who have not experienced illness and "old age," these aspects of lived experience may be terrifying. Ian James Kidd makes a helpful distinction between *pathophobic* and *pathophilic* attitudes toward illness (Carel 2018, 12). *Pathophobic* attitudes are characterized by fearing illness and wanting to avoid it at all costs. From the Western medical lens, illness is a problem to be solved or "cured." *Pathophilic* attitudes, by contrast, view illness as potentially edifying, positive, purifying, and instructive.

Pathophobic attitudes can also result in stigma, and despite being unfounded, result in oppressive circumstances for people with disabilities (which will be discussed in more detail in chapter 3). Stigma and psychological oppression lead to self-alienation and a depreciated sense of self (Bartky 2005). For example, stigma can affect how our loved ones view us, and in turn how we conceive of ourselves. Similarly, we may be perceived by strangers through a negative lens as well.

Enhancements and the (More) Exploitable Body

What is evident is that aging and illness are viewed from the perspective of the labor market with a *pathophobic attitude*. From the individual experiencing stigma, this can be costly in terms of social relations, job prospects, income, and support networks (Carel 2018, 71). As a boundary category, disability, including illness and aging, separates those allocated to either a work based or needs based system of distribution (Russell 2019, 4). Those who could be "cured," were essential to an exploitable workforce. By contrast, those who could not be cured were segregated and shoved out of the mainstream workforce. As Russell notes, disability "is a social creation which defines who is offered a job and who is not; what it means varies with the level of economic activity" (ibid.). Genetic technologies such as enhancements are merely tools to help avoid aging and illness in order to prevent reduction or expulsion from the labor market.

Enhancements view aging and illness, as well as disability, through a *pathophobic* lens. From this perspective, the ideological interests of capital, which views humans as either producers or consumers, becomes clearer when one considers how one might lengthen one's years of "youth" or strengthen one's "health" (Nocella II, Bentley, and Duncan 2012, XVI; cf. Nocella II, George, and Lupinacci 2019). As people with disabilities already know, they are viewed as either lower producers or limited consumers, unless the consumer has significant purchasing power. The only area in which people with disabilities are typically significant consumers concerns medicine, technology, and other health-related goods: "[c]onsumption supports the engines of production because people have to work in order to buy" what they desire (ibid., 17; Marcuse 1969). Stretching the length of years for an exploitable body, or further, offering a product for consumption that promises youth and health are both within the ideological aims of capital.

Finally, those who aim to cure or correct aging, illness, and disabilities are aiming at a moving target. Historically, disability, including aging and illness, has not been a static definition. Instead, it is dynamic because it is linked to the needs of capital accumulation. In this way, the category of disability is similar to the category of "race." As Charles Mills notes:

> Rather, racial identities are seen as historically constructed over time in response to historically variant political projects and shifting constellations of economic interests. But this "construction" is not arbitrary and purely "discursive"; it is motivated, materially enabled, and objectively rationally intelligible, contra postmodernism, precisely by the facts of a particular "metanarrative," the expansion of colonial capitalism. (2003, 128)

From the expansion of colonial capitalism, the idea of "the Black *worker*" was "introduced into the modern world system" (Robinson 1983, 199). As Cedric J. Robinson extracts from W.E. Dubois's first chapter of *Black Reconstruction*, "it was not as *slaves* that one could come to an understanding of the significance that these Black men, women, and children had for American development. It was as *labor*" (ibid.; see also Robinson 2019, chapter 17).

The category of disability shifts also in response to the demands of capital. For example, consider when there is a governmental crisis which results in a narrowing of the definition of disability, further leading to entitlement cuts (cf. Russell 2019, 6). Thus, capitalism has resulted in contradictory outcomes for people with disabilities: on the one hand, better medical technology to increase lifespan, slow aging, etc., and on the other hand, rigid and arbitrary diagnostic categories in incarceration and oppressive institutions (ibid.).

The mode of production is more accommodating for people with milder disabilities (Nocella II, Bentley, and Duncan 2012, 224). Historically, people with disabilities could still assist in peasant households. With the emergences of European imperialism, those who could not contribute to capital accumulation were disenfranchised and marginalized from the market. Marx (1867) commented upon this historical process in the mode of production in *Capital*:

> The proletariats created by the breaking-up of the bands of feudal retainers and by the forcible expropriation of people from the soil, this free and rightless proletariat could not possibly be absorbed by the nascent manufacturers as fast as it was thrown upon the world. On the other hand, these men, suddenly dragged from their accustomed mode of life could not immediately adapt themselves to the discipline of their new condition. They were turned in massive quantities into beggars, robbers and vagabonds, partly from inclination in most cases under the force of circumstances. (896; see also Nocella II, Bentley, and Duncan 2012, 226)

As a result, many were left destitute or placed in prisons, almshouses, and asylums. Those who fell into the three D's—defectives, delinquents, and dependents—were then the new targets for capitalistic eugenics (Nocella II, Bentley, and Duncan 2012, 229). Calls for sterilization (*Buck v. Bell*) and institutionalized abuse (e.g., Willowbrook) were common practice until the1970s, when Congress passed the Developmental Disabilities Act of 1975, which set for a Bill of Rights for people with developmental disabilities and established standards for habilitation programs (ibid., 233). Margaret Price, for example, has noticed the rhetorical strategy of juxtaposition which journalists use to link violent behavior to pathology, deviance, and "madness": their "choices of language and style create an overdetermined narrative of madness leading inexorably to a violent explosion" (2011, 146). It was not until Congress enacted the Education for all Handicapped Children Act, and the Civil Rights of Institutionalized Persons Act that certain protections were in place.

The effects of commoditization lead to the oppression of people with disabilities and is part of a systemic pattern of the logic of capital.[6] With the privileging of commodities and property over relationships, the result is a systemic depersonalization (Nocella II, Bentley, and Duncan 2012, 62). From the perspective of capital, people with disabilities have less exploitable productive value and cost more (ibid., 66).

The standardization, or normalcy, of labor is what capitalism wants. This leads to the promotion of controllable and normalized bodies for labor. Just as genetically modified food has become normalized, the ease with which genetically modified bodies might be "normalized" could be the same.

Anticipating the future, who is to say that those who are enhanced will not be more exploitable by those who own the means of production? Changing one's body will not entail a shift in wealth and private property. The basis for the oppression of people with disabilities is that it is easier to deny accommodations, which results in an exclusion from the workforce (Russell 2019, 14). From a Marxist perspective, disability can be reconceived as the outcome of the political economy (ibid., 2). While many disability scholars and activists have criticized the negative and prejudicial attitudes toward disability within the enhancement debate, even if those attitudes were corrected, the mode of production and concrete social relations that capitalism has created will still produce disabling barriers, exclusion and inequality. It is thus important to address how capitalism and the oppression of bodies may interact with future enhancement.

CAPITALISM AND THE OPPRESSION OF DISABLED BODIES

In *Learning from my Daughter,* Kittay remarks that many feminists have fallen for the faulty belief in the project of the Enlightenment because it offered the choice of whether or not to have children; "it is probable that women's achievements would not have advanced as far as they have without the right and the means to make these procreative choices" (2019, 66). Reproductive technologies, such as birth control pills, in vitro fertilization and genetic screening, have revolutionized relationships, family arrangements, and gender roles. These scientific advances have developed along with the changes in social relations and family arrangements. The hope is that genetic enhancements will expand upon this vision.

I am not confident that it will. Rather, it may exacerbate the problems of domination and oppression, which are already present within a capitalist society (Slorach 2016). Disability theorists and feminists have long cautioned against capitalism's tendency to be atomistic and to foster bureaucratic domination (e.g., Hall 2017 and Russell 2019). They, like me, wish to formulate a vision of social relations free from domination and oppression.

Specifically, the topic that has been overlooked in this debate is the role of oppression, and in this section, I wish to draw attention to the various ways that oppression may operate with enhanced capacities for an exploitable body. In "The Five Faces of Oppression," Young develops the five ways oppression operates in its systemic and structural forms (1990c, 40–41; cf. 1990a). Her account of the five faces illuminates how oppression makes up much of our social experience and is latent within major economic, political, and cultural institutions.

Young distinguishes between oppression and discrimination. "Discrimination" refers to the "conscious actions and policies by which members of a group are excluded from institutions or confined to inferior positions" (ibid.). Young articulates how oppression differs as follows:

> [t]he concept [of oppression] names the vast and deep injustices some groups suffer as a consequence of frequently unconscious assumptions and reactions of well-meaning people in ordinary interactions, media and cultural stereotypes, and structural features of bureaucratic hierarchy and market mechanisms, in short, the normal ongoing processes of everyday life. (1990c, 41)

Unlike discrimination, oppression concerns the invisible barriers which immobilize a particular group of people. An example is how the issues, experiences and embodiments of women with disabilities have been largely ignored (Schriempf 2001; cf. Lorde 1980). Oppression can be either structural or systemic and does not require that there is a correlated "oppressing" group for any specific group suffering from the effects. Instead, for every oppressed group there is a corresponding privileged group within society (Purcell 2014).

Exploitation

Young describes exploitation as the first face of oppression. As Charles Mills notes in *Black Rights, White Wrongs*, the concept of exploitation is "assumed to be tied to the labor theory of value, long repudiated not merely by mainstream economists but by even most contemporary Marxists" (2017, xix). Young, however, develops Marx's concept of exploitation to address issues of race and gender.

The Marxian theory of exploitation answers the paradox of capitalism: "when everyone is formally free, how can there be class domination? Why does there continue to be class distinction between the wealthy, who own the means of production, and the mass of people, who work for them" (Young 1990c, 48)? Capitalism systematically transfers the powers of the laborer to augment the power of the one who owns the means of production. Thus, the capitalist is able to maintain an "extractive power," which allows for the continual extraction of benefits from workers. This extractive power extends beyond the transfer of power: it also deprives workers of their sense of control and self-respect. It can accumulate passively for the capitalist, so that he may not even be aware that he is the beneficiary of such oppression (Purcell 2014).

In the case of racial exploitation, for example, there has been a long history of the "wrongful transfer of wealth and opportunities from people of color to whites (Mills 2017, xix). One example would be immigrants of color being paid less to do jobs than white workers would be (ibid., 128). A second example could include white candidates with less experience being preferred to Black candidates with superior credentials (ibid.). Who is to say that a similar divide would not occur between the enhanced and the unenhanced in the future?

Exploitation is relevant to the genetic enhancement debate regarding the advantages those who are enhanced may receive. While Alan Buchanan (2009), Bostrom, and other transhumanists have addressed the concern of distributed goods, their discussion has assumed that those who are enhanced will hold the power (e.g., Persson Savulescu 2012). Within a capitalist society, however, that may not be the case. In a capitalist society, the capitalist owns the means of production and extracts benefits from the laborers. It is quite possible that capitalists will seek to extract more from the enhanced laborers. Stated differently, while the enhanced may receive more opportunities for better jobs and better pay, it does not follow that they will possess the capital within a society. Thus, when we envision a future society it is necessary to include the strong likelihood of continued exploitation and the extraction of resources from enhanced bodies in the global marketplace.

This is not merely a concern raised by a dystopian fiction since we can already witness cases of this in our society. For example, we similarly witness this activity in the fertility business, which searches internationally for materials, such as wombs, for carrying children to be born. Moreover, there is a difference between the owner of a major league sports team and the player who is an employee. While it may be the case that enhanced athletes and enhanced researchers receive more job opportunities, it is not the case that the owners of those teams and laboratories will be enhanced as well. An account that does not include the possibility that the enhanced may become victims of exploitation due to their enhancements is incomplete. Simply stated, a theory that worries about inequitable stratification on account of genetic enhancement unhelpfully supposes that the envisioned society would meritocratically reward such enhancements.

Marginalization

Young's second face of oppression is marginalization, which may apply to those who are unenhanced. The "marginals" in society differ from those who are exploited, because marginals are the people the labor market does not employ. Many of those who are included in this group are certain racial minorities, elderly people, and people with disabilities (Young 2000, 169). A large proportion of the United States population are marginals: those who have been laid off from work and are struggling to regain employment, young people and people of color who cannot find first or second jobs, many single mothers with children, people with cognitive or physical disabilities, and Native Americans who live on reservations (Purcell 2014; cf. Asch 2002).

While some philosophers fear a genetically stratified society, we can recognize that this fear is based on the marginalization already present within a capitalistic society. For those parents who do not utilize these genetic technologies, or choose to have a child with a disability, the children are at risk of being born into a society in which the opportunities to secure resources may be limited or completely curtailed (Shickle 2000). The disadvantage results not from bodily diversity, but rather from the capitalistic structures which aim to accumulate more capital. Fair access to opportunities through government regulation will not ensure fair allocation. It could very well be the case that reasonable accommodation becomes viewed as an undue burden not worth the social and financial costs to employ the unenhanced who have become marginals.

Powerlessness

The worry about parental choices and marginalization leads to Young's third face of oppression, powerlessness. This form of oppression can be witnessed in the social division of labor between professionals and nonprofessionals. While exploitation and marginalization address class, powerlessness addresses status. Financial status and job security, especially for males, are considered to be an important determinant of psychological well-being (Oskrochi, Bani-Mustafa, and Oskrochi 2018). Powerlessness for Young, however, also includes social status: "[b]eing a professional entails occupying a status position that non-professionals lack, creating a condition of oppression that non-professionals suffer" (1990c, 56). Powerlessness describes the lack of work autonomy that nonprofessionals experience: they have little opportunity to exercise creativity or judgment in their work, have little technical experience or authority, express themselves "awkwardly," and "do not command respect" (ibid.; Purcell 2014).

In chapter 1, I addressed Licia Carlson's concern about experts as "gate-keepers" of knowledge. Similarly, within the enhancement debate we should consider the role that status will play in parental choices and whether the enhanced will receive an elevated social status within the new society. Even if equal opportunity is available for parents to elect enhancements for their future child, it is not necessarily the case that they will choose them. For some of the parents who refuse, it may be for religious reasons, while for others it may be due to a lack of scientific education. Furthermore, those who hold a lower status in society, and fear they lack the authority to make choices, may experience powerlessness when discussing genetic technologies with their physician. Parents may choose to enhance their children out of fear or insecurity because they do not feel they can exercise their reproductive liberty.

For example, Layla Al-Jader and Sharon Hopkins (2000) conducted a qualitative study to follow up on an antenatal screening program focused on fetal abnormalities for pregnant women in order to find out if the women had terminated the pregnancy or delivered the baby. What they found was that "[a]lthough all women wanted the choice to be screened, half of them were poorly informed to make decisions" (2000, 32). Moreover, "[w]omen who refused screening tended to be better educated and of higher social class" (ibid.). As in their study, powerlessness may play a similar role in future parental decisions regarding enhancements.

Marina Oshana (2006) notes that we should consider oppressive or external circumstances which can undermine an agent's autonomy. Oppressive environments which foster powerlessness and damaged emotional states, such as shame and low self-esteem, can curtail rather than embolden the reproductive liberty of future parents (Cudd 2006). Parents feeling powerless may lose self-trust because professionals such as physicians deem them of inferior worth due to their cultural background, lack of education, or limited financial resources (McLeod 2002).

Second, powerlessness may operate between the enhanced and unenhanced. Technology is not value-neutral, and as a result, those who are enhanced cognitively, physically, or morally may receive more esteem within a society. It may be the case that the unenhanced must "prove their worth" to receive similar respect. This is not so different from the problem of "Black Exceptionalism" within the United States: no matter how much wealth this person has acquired and how much status he or she has achieved, he or she will always be "marked" or "branded" by his or her color (Purcell 2014; cf. Wise 2009).

Cultural Imperialism

What is probably of most concern, though, in the current enhancement debate is the fourth face of oppression: cultural imperialism. While exploitation, marginalization, and powerlessness operate according to power, the fourth face of oppression operates according to power and recognition. In cultural imperialism, one dominant perspective renders another perspective as Other and, almost inevitably, as of less worth. For example, consider the demeaning or silencing of persons with disabilities in the enhancement debate among present philosophers. At its heart, cultural imperialism "consists in the universalization of one group's experience and culture, and its establishment as the norm" (Young 1990c, 59). Not cognizant of what they are doing, the members of the dominant group project their own experiences as representative of humanity as such (Purcell 2014). The result is that the dominate group treats the minority group as invisible. According to Young, "[t]hose living under cultural imperialism find themselves defined from the outside, positioned, and placed by a system of dominant meanings they experience as arising from elsewhere, from those with whom they do not identify, and who do not identify with them" (1990c, 59).

The oppressed group tries to express itself as a subculture because it desires the recognition of the dominant group within society but instead is marked as different or inferior. In this sense this consciousness is double. One's consciousness is double because one finds oneself defined by two cultures—a dominant one and a subordinate one (Purcell 2014). Thus, this is the injustice of cultural imperialism: "that the oppressed group's experience and interpretation of social life finds no expression that touches the dominant culture, while that same culture imposes on the opposed group its experience and interpretation of social life" (Young, 1990c, 60). An example of cultural imperialism is witnessed in the oppression of Deaf culture, which constitutes part of one's social identity (Dolnick 1993). Yet, this culture is almost always silenced, ignored, or unrecognized by a hearing community, which dominates the United States (or any other country).

Cultural imperialism is already at play within the enhancement debate. The tension arises from the perspective from which disability is viewed. On the one hand, some view disability as an atypical functioning of the body, and therefore it is naturally and immutably disadvantageous. On the other hand, others have argued that the issue is not the body, but rather the unjust social practice that rewards the "species-typical functioning" of those who are able-bodied and hinders those who are considered "atypical."

For example, some philosophers have argued that parents have a moral obligation to free one's future child of diseases and disabilities, and not to do so is an injustice (e.g., Fabre 2006). The bias at work here is ableism.[7] By not taking into perspective parents with disabilities in the genetic enhancement debate, philosophers enact a form of cultural imperialism.

A form of cultural imperialism which occurs at the individual level is silencing. Silencing as an epistemic injustice occurs when a person has been wronged in her capacity as an epistemic subject. Testimonial injustice identifies when a subject is wronged in her capacity as a subject of knowledge, and usually occurs when a hearer gives less credence to the speaker even though the speaker is supported by evidence. Hermeneutical injustice, by contrast, concerns wronging the subject in her capacity as a subject of understanding. This occurs when a subject's experiences are obscured from an individual and collective understanding, usually because of exclusionary practices. Both testimonial and hermeneutical injustice follow from prejudice (Fricker 2007).

For example, Carel notes that an epistemic injustice can occur in the case of illness. This can occur for testimonial injustice in which characteristics are attributed to those who are ill, such as cognitive unreliability and emotional instability. These attributions undermine the credibility of ill people's testimonies (Carel 2018, 11). Ill persons can also suffer hermeneutical injustice when aspects of the experience of illness are difficult to understand and communicate. In both ways ill persons can be denied in their capacity as knowers (ibid.).

If the testimonies of those living with disabilities are not included in the genetic enhancement debate, then the concerns raised are incomplete and an epistemic injustice has occurred (cf. Lackey 2020).[8] Systematically discounting the perspective of people with disabilities due to their declared (or undeclared) disability status wrongs them in their role as knowers. Further, they are discounted or excluded from the community of inquirers (cf. Longino 2001). The controversy for genetically eliminating a disability, rather than altering the social and environmental situations to provide accessibility and accommodation, is thus an example of cultural imperialism.

Young herself describes the discrimination that people with disabilities face as sometimes being a form of cultural oppression, in her essay "Social Movements and the Politics of Difference," but then at other times in that essay she describes those with disabilities as "unhealthy" and thus different from "normal" pregnant women (1990e, 175). Likewise, for the status of work, she argues that "[t]he variability of condition of people with disabilities is huge, however, and many of those brought together under this label have nothing at all in common in the way of experience, culture, or identity" (ibid., 171). Within the enhancement debate there has been confusion concerning aging, illness, and disability. Since many philosophers hope that enhancement technologies could be a "cure," there have been inaccurate generalizations concerning the sample set of disabilities and diseases considered. While this may be applicable in cases such as cancer or Alzheimer's disease, it is not the case for all disabilities.

Violence

The final face of oppression concerns violence in systemic and legal (though not just) forms. In this case, the oppressed group knows that they must fear violent, unprovoked attacks at random to their persons and property with the motive to "damage, humiliate, or destroy a person" (Young 1990c, 62). An example of this would be that any woman, because she is a member of the group "women," has a reason to fear rape. Intersectional concerns can raise one's susceptibility to violence: "disabled women are raped and abused at a rate more than twice that of nondisabled women" (Davis 2000, 332). People with disabilities are targets for such violence both in public places and in private institutions or group homes. They are targets for verbal, physical, and sexual abuse. In general, much of this kind of violence surfaces in the form of hate crimes, hate speech, or psychological violence and often goes unpunished within a society. According to Kittay, the "[w]idespread neglect and abuse of some of the most vulnerable among us" still continues (2002a, 290). How do we ensure that this neglect and abuse will not happen between the enhanced and the unenhanced?

The worry for genetic enhancement is whether violence would be a factor for a genetically stratified society. While violence against particular members of society would most likely continue in the case of enhanced physical or cognitive abilities, the question is whether it would continue with a morally enhanced society. This is part of the fear animating Persson and Savulescu's push to require moral enhancement (Persson and Savulescu 2012, 2014; Savulescu 2006). Buchanan et al. offer a solution for how future violence could be prevented, or at least, reduced. They give the analogy of criminal law. Insofar as law is justified to used coercive social means, so too society might "in the future attempt genetic interventions to do so as well" (2000, 173). The writers acknowledge that while violent behavior is not a disease, there are genetic factors that predispose individuals to violence. Reducing this disposition for violence while promoting cooperative behavior and altruistic concern through genetic intervention would benefit society as a whole (ibid.).

Sparrow (2014b) contends, though, that a society-wide program of genetic interventions for the sake of moral enhancement would be guilty of moral perfectionism. Even if a society-wide program for the enhancement of certain moral traits such as altruism or kindness were endorsed, then they would have to be universally required. This requirement, however, would violate moral pluralism, individual autonomy, and reproductive liberty. Simply stated, the only way violence within a genetically stratified society would not be likely to continue is in the case of a global endorsement of positive eugenics.[9]

While a person may experience any one of Young's faces individually, it is more likely that he or she will experience them in an overlapping effect. Because distributive theories of social justice regard individuals "as primarily possessors and consumers of goods to a wider context that also includes action, decisions about action, and provision of the means to develop and exercise capacities," Young argues that one should instead begin with the concepts of domination and oppression as a starting point (1990a, 16). The distributive paradigm conceptualizes social justice as a morally proper distribution of social benefits and burdens, including wealth, income, and material resources, among members of society. It also includes nonmaterial goods such as rights, opportunity, power, and self-respect. The fault here is that the distributive paradigm conceives of social justice and distribution as coextensive concepts (Francis and Silvers 2010; Silvers 2007). Those within the enhancement debate have taken on the same limited perspective.[10]

Capital unchecked leads to the destruction of human variation and diversity (cf. Mills 2003, 165–66). Yet, we have to remember that nature itself "does not normalize" (Nocella II, Bentley, and Duncan 2012, 98). Various abilities should be respected as part of a rich and holistic ecosystem. To account for the distribution of power and oppression, then, a different account of justice is needed. My proposal is that conceiving of justice as solidarity succeeds in this aim by turning to the virtue of respect for bodily diversity.

THE VIRTUE OF RESPECT FOR BODILY DIVERSITY

Rosemarie Garland-Thomson (2001, 2005) argues for an embrace of human variation and diversity. This embrace of diversity should guide feminist approaches in bioethics. Similarly, transnational feminist approaches help in taking a stance against a capitalist and globalist framework (Nocella II, Bentley, and Duncan 2012, 99; Fraser 2003). The virtue of respect for bodily diversity understands disabilities as conditions falling on a spectrum of physical or mental difference (e.g., Badenoch 2008; Armstrong 2011).

In this sense, one conceives of bodily difference in relation to one's society: for those who fall on the particular end of the spectrum it might be more of an impediment for living together in one's community because of the physical structures and social practices designed for the average members of that society. We must remember that neither disability nor disease is intrinsically an impediment to one's well-being.[11] Rather, it is the relation of living with a disability to one's society that can be an instrumental impediment to one's well-being if one is in the minority status on the continuous spectrum. Thus, when conceiving of genetic engineering regarding treatment and enhancement, one must realize that human embodiment cannot be de-linked from societal and environmental factors.

Two Features of Respect for Bodily Diversity

Respect for bodily diversity has two key features which challenge the labor market to think differently about the extraction of profit from bodies. First, respect for bodily diversity breaks down the fear of difference insofar as it provides the opportunity for those who are able-bodied to interact act with those who have disabilities and "therefore benefit from directly experiencing the humanity of persons who happen to be different in terms of the statistically normal ranges of ability for our species" (Woodcock 2009, 265).

To break down fears about alternative sources of the good requires a relational process of exchange. The process of forming a conception of the good itself in practice is relational (Francis and Silvers 2010, 249). The relational process of connecting with others builds trust and inclusiveness (ibid., 250). And this trust involves vulnerability (Baier 1986, 235). Diversity is an effective way to engage with our own vulnerability and to break down fears and prejudice. As a practice, it requires moral respect. The best expression of moral respect is "the willingness to listen to others express their needs and perspectives" but it does not require that we are able to imagine ourselves in their position (Young 1997, 347).

Second, respect for bodily diversity preserves benefits linked to disability. It is "beneficial to preserve diversity of human ability because characteristics that may seem wholly detrimental can turn out to be associated with other traits that provide valuable contributions to either the possessor or the broader community" (Woodcock 2009, 268). In some cases, it may not be possible to de-link the traits that are connected, such as bipolar disorder and creativity or autism and being a savant.

For example, while single-gene diseases are caused by a single malfunctioning allele, this is not the case with most traits or conditions under discussion in the genetic engineering debate. As Joseph Alper notes, because of gene-environment interactions, "gene A might result in a more favorable trait in environment 1 than would gene B in environment 2, but A in environment 2 might be less favorable than B in environment 1" (2002, 22; see also, Abberly 1987). Alper gives the following example to illustrate this difference in favorability with regard to teaching the method of reading: "Pupils with one genotype (the genetic makeup of an organism) will learn to read more easily using one pedagogic method, while pupils with a different genotype may do better with a different method" (ibid.). The lack of recognition of human variation follows from stigma about disability.

For debates concerning genetic technologies, we must remember that one's physical and psychological factors cannot be separated from one's environment, history, and social relations. This interconnection creates the possibility for the development of interventions for people with disabilities that are not guided by the normalization of their atypical or impaired functions.

Application to Facing Aging and Illness

So, what might the virtue of respect for bodily diversity be like in application to the lived experience of aging and illness? It requires a shift in perspective from *what* to *how*. Thinking creatively and valuing diversity, the answer to the lived experience of being unable to is to ask a different question: how might I do this or be this differently?

In the case of illness, Carel argues that this shift in perspective begins with recognizing that it is possible to experience wellness with illness. There are known to be three elements to help coping with illness and adversity: resilience, gratitude, and intimacy (Carel 2018, 13; Gilbert 2006). Experiencing illness does not entail that one cannot experience wellness. It is possible to experience wellness within illness. Carel notes that phenomenology calls attention to the implicit structures of experience, which may reveal our conditioned embodied existence. It may also indicate how we can be oblivious to kinds of embodied certainty and freedom that make it difficult to empathize with radically different forms of embodiment (ibid.). According to Carel, an ethics is integral to the phenomenological method: "a humbling recognition that our thought, experience, and activity fundamentally presuppose a way of being in the world, tacit bodily certainty, a sense of reality, and other features that are often taken for granted . . ." (ibid.).

Similarly, May Sarton argues that in the case of aging, we need to remember that the "concept of normality is a fictional ideal" (2007, 183). Aging in essence is what disproves the strict opposition between ability and disability in the first place; rather disability intersects with aging and illness "in an exploration of how bodies become culturally meaningful" (ibid.). To illustrate this phenomenon, she gives the example of the situation facing the "older woman" in American society (ibid., 183–85). According to Sarton, the older woman faces incompatible stereotypes. On the one hand, she is expected "to 'defy' its visible signs" by using cosmetic surgery to remake herself in the image of the "prevailing feminine icon" and to "manage" her aging body through "diet, strenuous exercise, or surgery" (ibid., 184). On the other hand, she is supposed to demonstrate traits of strength, liberation and self-control. Yet, in each of these stereotypes, mainstream American feminism has ignored, patronized, exploited, and silenced those who identify as older women (ibid., 185). In response, Sarton notes that in aging that there is more of an opportunity to turn a quotidian chore into a way to "celebrate companionship, aesthetic pleasure," or enjoy a "good meal or book" because one is no longer plagued by what one "ought to be doing" (ibid., 185–86).

In order to illustrate this change in perspective from *what* to *how* for aging and illness, it might be helpful to consider how Deaf culture has creatively cripped space and time. In music, if we consider the human perception of timing through rhythm and emotion, then we need to consider music interpretation in bodily variation. Amber Galloway-Gallego is one of many American Sign Language interpreters who specializes in signing for music concerts.[12] She has interpreted over four hundred artists in many genres, including hip-hop. Because ASL is its own language that is different from English, Galloway-Gallego must translate the various song lyrics into grammatically correct ASL. She then adds her own interpretation of the song through gesture and facial expressions. In her design and interpretation of each song, Galloway-Gallego crips time by communicating rhythm and emotion without sound. In this way, her creativity shifts a falsely assumed loss into a gain. When we think of bodily diversity as a gain rather than as a loss, then we gain a creative, cultural, and experiential way of being in the world.

By cripping space, we can also see how respect for bodily diversity alters the how of lived experience. In the architectural design for Gallaudet University (a school for the Deaf), we can experience what is called DeafSpace. The philosophy of DeafSpace rests on five principles. The first principle is space and proximity because eye contact, facial expressions, and body language are important for American Sign Language. For architectural design, this includes avoiding long, rectangular tables or long rows of desks, typical of what one might find in a "traditional classroom."[13] Instead, they have round or horseshoe-shaped seating arrangements in order to emphasize facial expressions, eye contact, and bodily gesture.

The second principle is sensory reach, which refers to using your sense to read the environment. DeafSpace design aims to extend this reach such as viewing corridors through and between buildings, low-glare reflective surfaces to show the shadow of a person just outside of one's vision, and controlled vibrations, which can indicate footfalls of someone around the corner.

The third principle is mobility and proximity, which is for signed conversation while moving, i.e., "on the go." With this principle, moving conversations which allow for observing a companion's face or broad array of bodily gesture calls for ramps and wide, gently sloping stairs. Intersections become softer and repetitive landscape or architectural elements may be used to reinforce the navigation of paths to travel.

The fourth principle is light and color, specifically muted blues and greens which are easier on a signer's eyes. For Gallaudet, this includes soft and diffuse lighting, which avoids dimness, glare, and backlighting among other visual distractions. Finally, there is the fifth principle—acoustics. The aim here is for an acoustically quiet space, because hearing aids and cochlear implants amplify sounds. Similarly, the humming of machines or technology as well as echoes can be distracting.

The value of diversity, then, is a public good that ought to be promoted to an extent (Woodcock 2009, 271). People with disabilities are contributing a "valuable social good by bringing unique characteristics to the moral community" (ibid., 272). Leslie Francis and Anita Silvers have proposed a way to construct ideas of the good with the respect and inclusion of people with disabilities (2006, 2009). Disability theorists have maintained that people with significant cognitive and communicative impairments can and do choose among alternatives. The issue is whether those choices are formulated in a way that they understand, and in some cases, using technologies to aid in this communication can be helpful (Francis 2009; Ferguson 2001). If one speculates that living a life without certain capacities may be devastating, then one's speculative judgment may be made in error (Asch and Wasserman 2005).

The virtue of respect for bodily diversity, then, helps us to recognize that genetic technologies cannot help us retain the activities that we fear aging and illness with take from us. To change our bodies rather than our perspectives will only leave us more exploitable to the various faces of oppression by capitalism. Instead, by shifting our perspective to valuing diverse bodies, we open ourselves to understanding that aging and illness challenge the *how* of our activities in lived experience, but not necessarily the *what* of those activities.

CONCLUDING THOUGHTS

As was evident in the discussion with my friends at the Halloween party, the choice I might have to make never concerned which enhancement I would want; the choice concerned which future child I would want. Was that a choice I wanted to make? And further, how would I make my choice? Most likely, I would have been advised to terminate the pregnancy of the fetus that had been deemed "less healthy" based on a trait or traits, which would have been based on a construction of "normalcy."

The exclusion of bodily diversity within society along with a reinforcement of normalcy is a problem which allows capital to exclude people with disabilities from economic life (Nocella II, Bentley, and Duncan 2012, 17). Because disability is a "big tent" as Kittay notes, meaning that people have different and unique needs, the terms that disability activists use often vary. This variance is done in order to resist capital's tendency to reduce humanity to the roles of producer and consumer (ibid.; Slorach 2016; Braverman 1974).

The task of Marxist praxis is to identify structural injustices and seek to rectify them through action. The liberal approaches suggested by transhumanists and other enhancement proponents fail "to expose either the way societies organized for the production of the material conditions of its existence or the mode of production plays the chief causal role in determining oppressive social outcomes" (Russell 2019, 12). For Marx, the "principle motive for historical change is the struggle among social classes over their corresponding shares in the harvest of production" and it is this situation which blocks the "class of disabled persons from the right to enter the labor force" or to place the disabled body within the political economy of capitalism (ibid.). This plays out in five faces of oppression. How will this not continue with enhancements?

I think enhancements are more likely to exacerbate domination and oppression rather than mollify the problems we already have. In this next chapter, I address death as a face of vulnerability and turn to the practice of storytelling for the virtue of relational authenticity. As Rachel Haliburton notes, those with illness are "wounded storytellers" who need to be able to "tell their illness narratives in order to make sense of what is happening to them" (2014, 239). In chapter three, then, I investigate the role our life story plays in helping us face our own finitude in the particular case of terminal illness and why genetic enhancements cannot alleviate our fear of mortality.

NOTES

1. For example, Gardner thinks that the use of genetic technologies for genetic enhancement may lead to an empirical slippery slope. An empirical slippery slope occurs when relevant moral distinctions do not adequately influence our choices. In this way, Gardener argues, "we will roll down the slope even though we may know better" (1995, 67). At the basis of this slippery slope are the choices of competing parties: an individual is responsive to the choice made by another, and similarly, societies are responsive to the choices made by competing societies. Following this line of thought, genetic enhancement for reproductive purposes may become the rational choice in the future for competitive parents and societies (ibid.). Ladelle McWhorter summarizes the debate over genetic enhancement in the following way: that it "purports to be a deliberation about whether one should act in certain ways, such as (1) support (or, in the case of scientists, engage in) genetic research leading to the development of enhancement technology (and if one is a clinician whether one should then facilitate its application), (2) vote for candidates for public office who would in turn vote for government funding for such research and technological development or at least refuse to ban it, and, perhaps, (3) offer oneself or one's progeny as test subjects" (2009, 410). She argues that instead of considering the viewpoint of others, those writing in this debate often counter or oppress their perspectives (ibid., 411). In many

ways, they are guilty of what Miranda Fricker (2007) calls epistemic silencing, when the hearer silences the speaker by not acknowledging their perspective.

2. Adrienne Asch (1999) has argued that the practice of prenatal diagnosis stigmatizes people with disabilities. It confuses the trait with the future child. See also Asch and Getter (1996) and Asch and Fine (2009). Concerns for civil liberties for the unenhanced are raised by Wikler and others: "Liberty for the unenhanced would place a burden on their more clever fellow citizens, one that might not be fair to them" (2009, 350). For debates about civil liberties, disability and genetic enhancement, see Tonkens (2011), Baylis and Robert (2004), Giubilini and Sanyal (2015), and Lin and Allhoff (2006).

3. Carel follows SK Toombs' five losses that one can experience in illness: wholeness, certainty, control, freedom, and the familiar world. According to Carel, the loss of wholeness comes from the perception of bodily impairment—the body can no longer be taken for granted because it thwarts plans, impedes choices, and renders actions impossible (2018, 42). Further, the body is now experienced as other. The loss of certainty, by contrast, forces the individual to face their own vulnerability and give up the assumption that they are indestructible (ibid.). The loss of control happens when we realize that our belief in medical science and technology protect us is an illusion. The fourth kind of loss, i.e., the freedom to act, can be exhibited by a lack of knowledge for which medical treatment or advice to pursue (ibid., 43). The fifth kind of loss—the loss of the familiar world—arises from the "disharmony illness causes and its distinct mode of being in the world" (ibid.). The ill person can no longer perform certain daily activities in the same way one did before. It should be noted that not all of these losses are necessarily experienced by the ill person.

4. For an overview and critique of transhumanism, see Nicholas Agar (2007).

5. In the literature, sometimes transhumanism and posthumanism are used interchangeably. There is a difference. Transhumanism is often used by analytical ethicists in the enhancement debate whereas posthumanism is a movement more closely connected to the continental tradition of philosophy. Both movements reject the belief that humans hold a special status due to some factor such as "human dignity." Posthumanism aims to move "culturally beyond categorical dualities concerning ethical and ontological issues" (Sorgner 2015, 32). Members of the Frankfurt school of philosophy, e.g., Jürgen Habermas (2003), however, do not align with posthumanism.

6. Robert McRuer (2006) points out that homosexuality and disability share a pathologized past. As a result, the heterosexual, able-body operates as a non-identity in the natural order of things. This cultural shift in neoliberal capitalism has created "more global inequality and raw exploitation" (ibid., 3). Further, "hard bosses and exploitative conditions helped to secure both able body identities in the sense that such identities needed to be reproduced for the future" (ibid., 94).

7. Gregor Wolbring writes that disability studies use the term "ableism" to "question and highlight the expectations toward species-typical body abilities" such as the view that birds have the ability to fly based on their species, and the disablement, which includes "the prejudice and discrimination people experience whose body structure and ability functioning are seen as sub species-typical and therefore labeled as 'impaired,'" such as not having curb cuts for wheelchair users (2012, 295).

Various scholars have responded to the inadequacy of the Rawlsian framework and unjust treatment of people with disabilities (e.g., Nussbaum 2001). Sophia Wong (2007, 2010) has tried to rectify this framework. Some scholars, such as Eva Kittay (1999, 2002a, 2002b, 2009, 2011), Rebecca Dresser (1995), and Hilde Lindeman (2001), have chosen to address these perspectives through the relations of care and policy. Others, such as Brueggeman et al. (2001), David Mitchell and Sharon Snyder (2016), and Andrea Nicki (2001), have sought to bring to light the testimonies and perspectives of people living with disabilities. Finally, some philosophers, such as Christopher Riddle (2014a, 2014b), have sought to integrate the perspectives of people within the capabilities approach as a response to the inadequacy of the Rawlsian framework.

8. Jennifer Lackey (2020) notes that within the example of the criminal justice system, false confessions should not be given as much credibility. There are limits to the credibility a hearer should give to a speaker in specific instances. Similarly, in the enhancement debate, some people with disabilities may hold false beliefs about their own experiences. For example, Gupta (2010) points out that in India, women with disabilities are targets of discrimination and oppression. Many women fear becoming pregnant with a female fetus with an abnormality. If the fetus does have an abnormality and is female, the mother almost always elects to terminate the pregnancy.

9. The moral fault of positive eugenics was not that it relied on poor scientific understanding, but that it violated reproductive liberty (see chapter 4). If pro-enhancement philosophers such as Buchanan, Persson, Savulescu, and Bostrom still hold that moral enhancement is in the best interest of society and should be required against the will of the parents, then the charge of positive eugenics sticks.

10. To broaden this perspective, it is also necessary to consider the way power operates within society. What has been assumed by philosophers endorsing genetic enhancement is an implicit or explicit conception of power as "a kind of stuff possessed by individual agents in greater or lesser amounts" (Allen 2008, 31). Power should be understood as a relation rather than a thing. Further, their constructions conceive of power as a relation of atomistic persons in which one agent (the enhanced) exercises power over another (the unenhanced). Third, the cooperative framework proposed understands power as a fixed or static pattern of distribution rather than as a dynamic process. Finally, the fear of the enhanced dominating the unenhanced understands domination "as [a] function of the concentration of power in the hands of the few" and hopes to remedy this problem with a "redistribution of power that strives to put [more power] in the hands" of the many (Young 1990d, 158–59).

11. What is of confusion in this debate are two senses of disability. Ron Amundson and Shari Tresky (2007) draw a distinction between "conditional disadvantages of impairment (CDIs)" and "unconditional disadvantages of impairment (UDIs)." Conditional disadvantages of impairment are produced by the social context for people with impairments. By contrast, unconditional disadvantages of impairment are experienced irrespective of the social context (2007, 544). The conflation of these two terms has led to stigma and oppression within the debate. Unconditional disadvantages of impairment are not controversial. What is of concern for disability

scholars is biased disregard that treats conditional disadvantages and unconditional disadvantages as the same.

12. For information on Amber Galloway Gallego, please visit her website at https://www.ambergproductions.com/about

13. Information about DeafSpace architecture principles can be found here at https://archive.curbed.com/2016/3/2/11140210/gallaudet-deafspace-washington-dc

Chapter 3

Facing Death

Invulnerability and the "Cure"

INTRODUCTION

I had known Chivel for nine years when she "lost her battle" to brain cancer. Every year I would travel with my spouse and stay with his family in México. I could barely speak any Spanish and Tía Chivel could barely speak English. Yet, every journey I made she would welcome me into her home with open arms and tell me how much she loved me. Her home was always a lively place to stay with her husband Hugo and their three children Erica, Jessica, and Hugo Jr. (all my spouse's cousins). Early in the morning when the coffee was made and the birds were out, I would be awakened by music and Hugo cheerily singing. Over the years I was blessed with knowing Tía Chivel, I learned many lessons; of those lessons, she taught me not only the courage to face one's own death, but more importantly the courage to accept and to embrace our human fragility.

Facing our own death and the deaths of our loved ones point to the contingency of our life paths. We will never know how significant we were in this world after we pass on. And before we die, we all have the potential to suffer physically, psychologically, morally, and spiritually. There is a temptation to reach for a possible future in which we do not suffer and face death—and that temptation is not new (May 2017, 67).

According to Todd May in *A Fragile Life* (2017), the conception of the afterlife offers us a way around death, i.e., the temptation to reach for an end to anguish and an alleviation for death (ibid.). The philosophy we embrace is that a good life is available for us all. The question, then, is how can and should we live that good life? While philosophy can answer the question of should, it is genetic technologies that now offer the possibility of how we could live that good life: a life that one can control (ibid., 88).

For us today, genetic technologies might be the way to control the possibility of a good life for one's future child; moreover, they might extend the years of the good life all of us might have. If this is the case, should we not embrace them? What would possibly be the logic to make us reject the ability to ensure that our future children (and even us ourselves!) have a better life than we have currently? In other words, don't we owe it to our future progeny to give them a future better than ours?

In this chapter, I address the aim of genetic technologies to alleviate our fear of death. Many philosophers have proposed that health could be considered as a primary good to be engineered or enhanced. However, this desire for immortality, or at least invulnerability, I argue, is a form of idolatry in the Marxist sense. The aim of trying to prevent a vulnerable body is misguided. Rather than hide from our own mortality, it is better to have the courage to face it. This courage is found with relational authenticity, a virtue enacted with others through solidarity. To begin, I want to address this face of vulnerability before moving on to why some philosophers think we might be able to postpone it.

DEATH AS A FACE OF VULNERABILITY

A common feature of the human condition is to fear our own mortality. Some philosophers have argued that we fear death because it is a loss of life, and so a life that passes through all the stages of the life course is a life well lived. Others, such as Martin Heidegger (1962), have argued that facing one's finitude can provide an orientation to one's being-in-the world. Finally, specific philosophical traditions, such as Stoicism and Buddhism, offer counsel to focus on the present rather than concern ourselves with worry about our future demise. What these approaches offer is a kind of emotional invulnerability in the form of mental tranquility. Regardless of which approach one takes to this face of vulnerability, facing death frightens us in part because it forces us to rewrite our life story.

When we meditate on death, we still think of ourselves as present. As May writes, "we fear the state of being there but being dead" (2017, 101). Death reminds us of our own fragility in the world. The lived experience of facing death brings us back to our own time frame. It becomes a personal lens from which to view and judge the world.

Personally, death means the end of *my* future. It is the end of *my* possibilities. It is the end of *my* experiences, activities, and projects. And I have very little to no control over how my death affects others. The face of death frightens us because this is not how we thought our story would end.

The lived experience of facing death reveals to us that although we are the authors of our own life story, we are not fully in control of how our story will go. In this chapter, I want to limit my focus to the experience of being diagnosed with a terminal cancer to describe various features of facing death. In order to describe the lived experience of facing death, I think it is best to listen to the stories of those who have shared what it is like to reckon with this face of vulnerability. Finitude is a face that we will all encounter at one point, and yet, in Western society, it is the face we deny the existence of the most (Vig and Pearlman 2004).

The late philosopher Jeff Mason wrote that "birth and death are the bookends of our lives" (2011).[1] Death can bring direction to one's life and can give a different framework from which to consider the changes that life brings. According to Mason, the "old look back," while the "young look forward." What matters to us changes as we age. Death only informs us of these changes. For the young, mortality is only an idea; it is not yet a possibility. For the old, though, the reality of death does start to sink in.

At the age of sixty-seven, Mason received his diagnosis of stage IV inoperable lung cancer with bone metastases (Mason 2012). He said, at first, he felt nothing at all, and then he felt gratitude for the life he had lived. His own mortality had become a reality. Many have described the five stages of grief and loss by Elisabeth Kübler-Ross: denial, anger, bargaining, depression, and acceptance. For Mason, though, he said that he skipped straight to acceptance. After six months of living with lung cancer, he tried to understand what his death would amount to. He describes death as a shadow that one casts on a sunny day. It is there whether we deny it or worry about it. We just have to accept it.

The lived experience of facing death can only be described from the perspective of the living. Often, it is a concept that we avoid in our thoughts and actions. When we do think about death, it is usually from a detached point of view such as the sort afforded in a lecture hall. And when we think about, we do so from the position of ignorance.

The habit of ignoring death gives us the false impression of life's permanence rather than its fragility. Ignoring death encourages us to obsess over the minutiae of daily life. And yet meditating on death too much can lead us away from the joys of life. Facing death with courage means coming to terms with one's life: its significance, its meaning, and determining whether we led a life well lived. And this courage is enacted, in part, by the stories we tell.

Stories told by those who have terminal illness or other conditions facing death give voice to those who have been silenced within society. Rachel Haliburton talks about illness narratives, which are narratives that demonstrate the perspective of the sufferer (2014, 230; cf. Frank 1995). Illness narratives are not just simply a story about one's own experience, but an exploration of the lived experience more generally. Part of the need to tell one's story is to make sense of what is happening to oneself. Consider Vanessa's story as she describes her experience when she received her diagnosis of colorectal cancer at the age of twenty-seven.[2]

Vanessa went to urgent care because she was experiencing symptoms like a stomach flu that would not go away. After a series of tests, she received her diagnosis. At first, she felt denial and shock. She then describes how she went through the process of setting up procedures in an almost robotic like fashion.

Serious illness, as in the case Vanessa faces, can leave one "stranded, without a map and a destination" (Haliburton 2014, 239). Those experiencing serious illness act bravely by "confronting the illness, embracing the medical treatment and patiently waiting for the attention of physicians" (ibid.). The assumed order to our generally chaotic and uncontrollable lives has been lost.

In reacting to the future anticipation, Vanessa said she wanted to cry and wanted to joke. She wanted something to keep her mind off it. One of her treatments would be radiation, which would cause her not to have children unless she could possibly freeze her eggs. She said she felt blessed that the doctors moved quickly and that they gave her a prognosis of an eighty-ninety percent chance of survival. Because she was so young, doctors did not think to test her for cancer. Vanessa describes her situation by asking where she might have been in four to five years. She might have run out of time. She hopes that they remove it and that she lives for many more years.

Yet, hope was not Vanessa's first response. Instead, she said that when she first received the diagnosis, she felt hurt and felt that her body had betrayed her. She was scared, then angry, bitter, and then asking why. Hadn't she suffered enough? She had just lost her mother to cancer. And then she thought about her family. She was happy then that it was her and not them.

What Vanessa's story shows is that illness disrupts the order of how we think our life story is supposed to go. The challenge is to find meaning and create order in our life story in response. This can involve even thinking of one's past self as a different person.

The story of the lived experience of terminal illness, though, is a specific kind of illness narrative. Called "pathographies," these stories can be told by the individuals themselves or by those who are allies speaking in solidarity with them. *Pathography* is a "form of autobiography or biography that describes personal experiences of illness, treatment, and sometimes death" (Waples 2014, 155). For example, autopathographies are "personal narratives about illness or disability that contest cultural discourses stigmatizing the writer as abnormal, aberrant, or in some sense pathological" (ibid.). In autopathographies, the "moment of death is implied by the chronological narrative trajectory, inching ever closer as the pages grow slim toward the end of the text" (ibid., 156).

Yet, facing death, as in the case of terminal illness, does not just alter one's life story; instead, it rewrites it. Take for instance the story of Tamara O'Brien. An elite athlete, she was diagnosed with terminal cancer at the age of twenty-two.[3] She describes her first reaction as a state of confusion with how she was living her life. She had been in gymnastics since the age of two, and then made it to the Canadian National Team by age eleven. Her goal was to compete at the World Championships. Only weeks before she was set to compete did she notice a lump on the left side of her neck and then a lump under her chin. After medical tests, she was told she had melanoma in her lymph nodes, and they would need to do surgery in two days.

She said her reaction to this news was disbelief because she was set to compete at the World Championships. After a look from her doctor, Tamara recalls the realization she had that the championships were out and after the surgery, she would need to figure out what her next steps would be. She needed more surgeries, and yet they still could not remove all the melanoma. She had reached a stage IV diagnosis after having been "cut open five times." She was lost, felt defeated, exhausted, and in pain.

When she met with her oncologist, she describes how the realization of the situation affected her. The oncologist was listing off all the metastases in her liver, her bones, her lymph nodes, her groin, and her armpits. It was all unreal; they were trying to fix it, cure her. And then, she realized there was no more surgery that could be done. It was then that Tamara thought, "oh my God. I really have cancer. It was the worst day of my life" (CBC Docs 2019).

Facing death, as in the case of terminal illness, disrupts the linear structure of normative time in our overarching life story. In time, facing our own death intersects our perspective of our past and future. For the past, we are affected by what we have not yet done. For the future, we are affected by what we won't be able to do. And these two can come together in the realization of what we will miss when we are no longer here.

There are those that receive a diagnosis and then later are cured. And then there are those like Tamara and Jeff, who are not. After much reflection, this is what Tamara had to say on those who have survived cancer and consider it a gift: "I don't think cancer is a gift, but it is an awakening. So, it woke me up. It really woke me up" (ibid.). Before her diagnosis, Tamara considered herself strong because she had mastered gymnastics, could compete at an international level, and perform in front of a large crowd. As she described it,

> I thought I was that person and then . . . that girl had no idea. She had no idea what was about to happen. I feel bad for her . . . She didn't love herself. And she never considered herself beautiful. And she hated how she looked, you know? She never thought she would really amount to anything huger other than in her sport, and that's so sad to me . . . I live my life with meaning now (ibid.).

Choosing to live each day with meaning, Tamara said she was grateful for the moments that she has been given in her life. And she has much more confidence than she used to have. When asked whether she would trade not having cancer to go back to her life before, she said that she did not think that she would. She doesn't want to be that old girl. She has learned so much since then.

Tamara's story describes how facing death causes us to rewrite our life story. In her realization of her cancer being terminal, Tamara began to live a different life story than the one she had imagined and anticipated from years before. No longer focusing on winning World Championships, instead she sought to live each day with meaning and purpose. She reclaimed the authorship of her future. She changed, and so did the trajectory of her life story.

Facing death, though, does not just force us to rewrite the timeline of our life course; facing death can also alter our sense of space. What was once familiar becomes uncertain and even silencing. Audre Lorde captures this disruption of space by telling her story of living with cancer in both *The Cancer Journals* (1980) and *A Burst of Light* (1988).

In *The Cancer Journals* (1980), Lorde wrote her personal and political account of the lived experience of breast cancer. Here, she brought the privatized areas of shame, fear, and guilt into visibility. Her aim was always to fight against the scourge of silence. And the first step, she argues, is to become visible to each other.

As Lorde notes, silence, invisibility, and powerlessness go hand in hand (1980, 162). Prosthetics, cosmetic surgeries, and contemporary programs like Look Good Feel Better[4] silence and separate women from each other as "the embodied realities of breast cancer become subject to wide-scale social erasure" (ibid., 167).[5] For example, Lorde writes:

> Here we were, in the offices of one of the top breast cancer surgeons in New York City. Every woman there either had a breast removed, might have to have a breast removed, or was afraid of having to have a breast removed. And every woman there could have used a reminder that having one breast did not mean her life was over, nor that she was less a woman, nor that she was condemned to the use of a placebo in order to feel good about herself and the way she looked. Yet a woman who has one breast and refuses to hide that fact behind a pathetic puff of lambswool which has no relationship nor likeness to her own breasts, a woman who is attempting to come to terms with her changed landscape and changed timetable of life and with her own body and pain and beauty and strength, that woman is seen as a threat to the "morale" of a breast surgeon's office! (ibid., 59–60)

By contrast, men with wounds are considered warriors. Lorde argues that this is a call to action for women with breast cancer to be warriors, too. As the primary scourge of our time, women are at war with cancer. Lorde continues: "I refuse to have my scars hidden or trivialized behind lambswool or silicone gel. I refuse to be reduced in my own eyes or in the eyes of others from warrior to mere victim" (ibid., 61). Like a call to be warriors of the Amazon, Lorde refuses to change her body to make a "woman-phobic world more comfortable" (ibid.).

By 1984, the cancer had spread from her breast to her liver. In *A Burst of Light* (1988), Lorde wrote about her decision to forgo the usual treatments and instead pursue homeopathy and self-hypnosis. In reclaiming the authorship of her life story, Lorde chose to continue living her life, defining her purpose, and pursuing alternative treatments. She even planned a teaching trip to Europe.

The metaphor of travel and baggage she must leave behind mirrors her actual travels after her diagnosis. By the spring, she had lost almost fifty pounds. Yet, she was determined not to be silent, and to do the work of visibility and kinship across difference. For example, she taught in Germany, became involved in international communities of African Diaspora, and traveled to the word's first feminist book fair in London.

In 1992, she passed away in St. Croix, US Virgin Islands from complications. This was one year after becoming the first African American and first woman to be named New York State poet laureate. Lorde lived nearly ten years after her diagnosis, and took the African name Gamda Adisa, which means "Warrior: She Who Makes Her Meaning Known" (Bolaki 2019).

Stories like the ones above must both be told and heard, and thus every story has two sides: the *personal,* the experiences told by the person telling it, and the *social,* how the story is received, whether heard, distorted or transformed into something more acceptable to the listener (Haliburton 2014, 240). It is a moral wrong when those stories go unheard (ibid., 238).

Storytelling, then, creates ethical obligations for tellers to present a "story in as truthful a way as possible" and for listeners to "understand the truth" (ibid.). It is to treat one another with respect. When we don't listen to each other, when we disregard or dismiss or distort the voices of those trying to speak, we wound the *self.* We wound the selves of the tellers when we do not listen. This is the part of us "which necessarily experiences things subjectively but needs to have that experience recognized by others if it is to be fully validated as meaningful and significant" (ibid., 245).

The personal and social dimensions of storytelling come together in Lorde's writings. As impressive as her travels were, Lorde's true courage can be witnessed from the perspective of her inner journey. As she wrote, facing death conceals our prospects and reveals uncertainty. It is difficult to hold onto hope and faith in a "sunless place" (1988, 40–52). Drawing upon courage in her desperation, Lorde writes she must stay open and look at all her options carefully. And then, she says that she can broaden the definition of winning to the point where she cannot lose (ibid.; cf. 110–12).

The language of fighting a battle, "winning," "losing," and "beating" illness such as cancer is not uncommon (Lanier 2020, 272–73). I used this terminology at the beginning of this chapter. The appeal of genetic technologies, to use this language, is that they can help us "win." Unlike Lorde, we would no longer need to broaden the definition of winning.

Rather than face death, these technologies help us avoid or postpone this face of vulnerability. Genetic technologies, especially enhancements, offer a way to keep control of the narrative of our life story. If we control the timeline of our narrative, then we can postpone or even eliminate that future worry about our eventual finitude. Further, they offer a kind of physical invulnerability. In a word, they off a kind of "cure" for many of the disruptions which cause us to rewrite how our story will end.

ENHANCING FOR BORROWED TIME AND A FUTURE BETTER THAN OURS

Enhancing Health to Borrow Time

To ward off our own mortality, in our present day we often try to control our eating and exercising habits in order to keep disease and bodily injury at bay (May 2017, 89). Genetic enhancement technologies, however, might be able to offer more control over our susceptibility to disease and injury. They might be able to offer us overall enhanced health.

Similarly, some philosophers think of death as a harm because it is a loss of life (Mason 2011). This is because it curtails someone's progression through the life course. To harm someone is to make them worse than they otherwise would have been. Unless we are in tragic circumstances, then, death is considered worse than life. Having better health overall, which enhancement technologies might be able to provide, could ward off a premature or unfortunate death.

In a line, genetic enhancement technologies could help us keep control over the narrative of our life story. They could do so by providing a "cure" to many of life's inherited health conditions. For example, genetic therapies or enhancements could target conditions such as Huntington's, Tay Sachs, Hemophilia, Schizophrenia, Bipolar disorder and Down syndrome, which many philosophers think lead to lower well-being. The enhancement of health overall could lead to an increase in well-being.

If sometime in the future genetic enhancements could enhance our health and enable us to live longer lives, in many ways, we would be living on borrowed time. Conceiving of functional enhancements as primary, or all-purpose, goods, philosophers such as Walter Veit (2018) and Fritz Allhoff (2005), have argued that health among other capacities should be enhanced.[6] The point, these philosophers argue, is that these primary goods are expected to make life go better, and thus, the genetic technologies which enhanced these capacities should be utilized.

Many assume that living with health conditions leads to a poor quality of life, and so using genetic technologies would appear to be rational in these cases to increase one's future health and well-being. For example, John Harris maintains that while he thinks people with disabilities do have equal moral worth, it is better that parents avoid, and even morally wrong, to bring people with disabilities into the world if they know about the possibility and abortion is an available solution (Walsh 2010, 521).

This is because many philosophers who conceive of enhanced capacities as all-purpose goods make use of a maximalist view of well-being. Having disabilities or multiple health conditions would be impediments to a good life. Jeff McMahan, for example, summarizes the issue in the following way:

> A single disability may seem neutral because it can be compensated for by other abilities that develop to fulfill its functions. Blindness, for example, may be compensated for by the enhancement of other senses, particularly hearing. But if disabilities were individually entirely neutral, they ought also to be neutral in combination; but they are not. (2005, 96)

McMahan's claim is that disabilities are not neutral in combination because with each further disability it becomes harder to compensate for other disabilities. It is in this sense that functional capacities demonstrate their value as all-purpose goods: not enhancing just one capacity but enhancing many will make life go better overall.

The appeal to all-purpose goods and a maximalist view of well-being is problematic for several reasons. First, it is impossible for us to entirely control the process of human development from the "inside out" by focusing on the "jewels in the genome" to engineer or enhance cognitive, physical, or moral traits (Sikela 2006; Sparrow 2011). Most of our makeup is an interaction between our genetics, other genes, and our environment. Environmental forces play a strong role in whom we become as individuals. These arguments, in short, stray too far from the present scientific ability to be meaningful yet.

For example, medical science cannot guarantee, by genetic interventions alone, that if we enhance our health that we will be healthy. Further, science cannot guarantee that our lives would complete the stages of the life course. It is a fact that often, even with the best intentions, the results of our actions fail to succeed.

When we look to evolutionary biology, many viruses, such as the 1918 Influenza Pandemic and SARS-Cov-2 (i.e., the coronavirus Covid-19 pandemic), can cause a cytokine storm for those with robust immune systems. In these cases, one's healthiness works against one's biological system. This is unfortunately the result of adaptative evolution—not of humans, but of viruses and other diseases—that the enhancement enthusiasts have overlooked. Human biology is not the only criterion to consider when we envision an enhanced future. Our enhanced healthiness cannot be guaranteed medically.

This is because the aim of medical interventions is to target a specific health condition. For example, vaccines have probabilities against recognized viruses, which cause recognized harms. What is at stake in this argument is what counts as "healthy" in the first place. Certain conditions, such as strabismus, do not mean that the child is unhealthy. Appealing to health as a primary good is too abstract to ensure that we are actually healthy.

Second, who should determine which traits are valuable? As disability scholars have pointed out, the eugenics movement made this same mistake, and the goal of perfecting or cleansing the human gene pool is already misguided (Buchanan et al. 2000). Similarly, what about disabilities which fluctuate over time (Reeve 2009)? Thinking along these lines of control for human development and reducing who we are to genotypes undermines biological diversity (Asch, Gostin, and Johnson 2003).

The point is that it does not necessarily follow that having a disability is an impediment to living a good life. As Asch and Wasserman argue:

> [H]uman beings enjoy a fortunate redundancy in many of the capacities that are instrumental for, or constitutive of, valuable human goods and activities, from intimate relationships to rewarding work. Humans with a standard complement of senses and motor functions rarely use all of these functions in achieving such goods, and humans lacking those skills can use only some. But those are usually sufficient. (2005, 208)

If the purpose of enhancements is to increase one's overall well-being and life satisfaction, then one must also accept that there may be multiple ways to realize the good. While some individuals may not be able to experience rich aesthetic experiences, it does not mean that their life is worse overall or that it will require more luck to have a good life. The philosophers in support of these technologies who want to eliminate particular traits deemed "disadvantageous, [or an] abnormality, flaw, or defect" have brought impairment into being as a kind of thing, i.e., reifying the individual interchangeably with the trait (Tremain 2017, 162).

Yet, while one could argue that health calamities could happen at any point in a person's life and medical interventions can fail, this objection does not hit the target, namely that of health as a primary good. There is something else going on in our quest for invulnerability and our desire to control the narrative of our lives. It is the possibility of gaining control as an *ideal.*

This argument for health as a primary good relies upon a *ceterus paribus* ideal. The concern is simply—now that we have the technology to ensure better health for our future children and possibly ourselves, should we not use this technology to do so? Genetic interventions have been designed to "even out" the biological challenges that once were left to chance or luck. We would be foolish—no unjust!—not to utilize them. The choice has never been about existentially cheating our future children out of autonomous decisions. It has been about existentially cheating them—due to omission—out of a fair deck of cards in the game of life. Why would we want to "stack the deck" against them? People like Tamara were cheated out of a fair deck. Don't we have an obligation to provide a future better than ours?

To answer this question, though, requires a different sort of response. It requires that we talk about the possibility of granting our children a future better than ours. In order to address this matter, I want to begin with a story close to my heart.

A Future Better Than Ours

My friend Kaitlin was told her four-year-old son Jacob had leukemia. When I heard the news, I wanted to throw up. As a nurse, Kaitlin knew what lay ahead for her son. With courage, she and her family joined together around their son, getting him the medical treatments they could. Jacob "beat cancer"—only to have it come back four more times over a period of fourteen years. Right now, he just completed treatments, and graduated early from high school.[7]

There are few greater assaults on us than the loss of a loved one, especially one's child. Genetic technologies, and in particular genetic enhancements, cannot protect us against the possibility that we will experience grief. Vulnerability, including losing someone we love, is a part of the human condition. The grief we experience is for our own sake and for the sake of the one lost. As May notes, our loved ones are woven into our lives, and are central to what makes our lives meaningful. Losing a loved one is also losing a part of who I have been and who I can be in the future. It is having "a piece of what makes my life meaningful shorn away from me" (May 2017, 125). Losing a loved one also forces us to rewrite our life story.

There is no substitution for the one that is lost. It does not mean that I cannot recover or develop new relations with others. It will result in a different type of meaningfulness. To try to mirror the previous relationship is to make lightness of the person who was lost. Grief, by contrast, marks a recognition of the death of someone about whom we care. It does not make sense to recognize a loss and not feel it. Electing to use genetic technologies to prevent the possible loss and grief we can experience is to disregard the vulnerability and the interdependence of the human condition.

If enhancements could protect our loved ones, or even cure them, should we not pursue them? If we have the possibility to provide the "best life possible" or a future for our children that is better than ours, should we not take it (e.g., Savulescu 2007, 2009b)? Are we not obligated to provide a better future to our children? And if we do not, then, are we stacking the deck against them by doing nothing?

In philosophical language, what I am asking is if we "opt out" of using genetic enhancement technologies, and thereby commit an omission by doing nothing, of what have we deprived our future children? The obligation to provide a better future assumes that we have harmed our future children by deprivation. And this assumption has an origin in classical abortion debates.

In these debates, one argument that stands out is the *Future Like Ours (FLO) argument,* which makes sense of a deprivation account of harm. The philosopher most closely aligned with this argument is Don Marquis. For FLO, Don Marquis (1989) argues that killing is wrong because of the effect on the victim. The greatest loss one can suffer is the loss of one's life. Killing is a harm because it deprives us of all the experiences, activities, projects and enjoyments that would have made up our future. In other words, if I am killed, then I am deprived of all the value of my future. I am deprived of a future like ours (Steinbock 2011, 65).

In a later article, Marquis says that a future of value contains a good life and is a life worth living. It includes "all those experiences and projects that will make our future lives worth living if we have the opportunity to live them" (1995, 1). A good life includes "friendships, loves, absorption in various projects, aesthetic experiences, identification with larger causes seen as valuable, such as one's team winning a victory, and physical pleasures" (Steinbock 2011, 65).

For Marquis, "the value of one's present life is irrelevant" (ibid.). What is essential about having a future like ours is that one's future is one's own, and this is the future one is deprived of (ibid., 66). It is my personal future, and because it is *mine*, it concerns an issue of ownership. The harm done to a fetus by abortion is that it is deprived of a future like ours. This is why FLO makes sense of a deprivation account of harm.

Marquis's FLO account provides an alternative way for thinking about the lived experience of facing death[8] and whether we have an obligation to use genetic technologies to give our children a future better than ours. When philosophers think about death, according to Mason, what is strange is that they try to do so from an impersonal or ideal stance. In doing so, the concept of death is empty, has no use or content, and has no object. It has no subjective meaning.

From this perspective, it would be reasonable to assume that enhancing capacities as primary goods, such as health, would lead to an increase in well-being. In this sense, the concepts of death, health, and well-being remain in the abstract. What is strange about this impersonal stance, though, is that it does not hit the target of why we fear the face of death: we fear death because it is *my* death or because it is the death of *my* loved ones. We fear the lived experience of death because it is personal, concrete, and intimate.

In philosophical language, then, the trouble I think that construing these concepts in the abstract runs into is the problem of vagueness. What I mean is that if we consider some set of including or excluding qualities X, it could turn out that these concepts are just an empty set. The insight I draw from Marquis's FLO is that our lived experience is personal. Facing death from the perspective of lived experience means meditating on the end of my future, my possibilities, and my experiences, activities and projects. Facing death means that I as the author of my own story must grapple with how my story will come to an end. Because my life story is *mine,* then it, too, is an issue of ownership.

Moreover, in a similar vein, making an argument that we must give our children "the best life possible" or a future better than ours is argued from a detached point of view. As Iris Young (1990a) pointed out, abstraction to the independent and universal, without presuppositions, makes one's theory too impoverished to be useful for the evaluation of someone's actual well-being along with the social situations, institutions and practices which uphold it. To be useful, it must contain some premise or premises about social life, which are drawn from social contexts. When philosophers abstract human enhancement from the particular circumstances of social life which give rise to the concrete claims of theories of well-being, they falsely assume that one can distinguish objective claims of well-being from the socially specific claims given for a life well lived.

A case in point is the concept of the "cure."[9] As I have pointed out, the language we use concerning terminal illness is metaphorical. It is an hypothetical attempt to fill in the concrete details of a crisis we are facing. The concept of a "cure" is an ideal we hope for and pray for. And yet, medical technologies can only provide treatments for certain conditions in the concrete. To conceive of genetic enhancement technologies as providing a "cure" in the abstract is to assign these tools with special protective powers.

Whether health as an all-purpose good is enhanced, or one argues that we have an obligation to provide a future better than ours, what seems to have been assumed in the enhancement debate is that these technologies have the power to limit our vulnerability, or even eventually, grant us physical invulnerability from many diseases, health conditions, and impairments. But, I have to ask, why are we ascribing such mystical power to scientific tools we have created in the first place?

STIGMA AND THE IDOLATRY OF THE "CURE"

Idolatry and the Cure

I think the answer to this question lies in the origin of Marx's conception of alienation: idolatry.[10] Erich Fromm contextualizes the broadening of the term and argues that the concept of alienation can be traced back to the Old Testament account of idolatry:

> The essence of what the prophets call "idolatry" is not that man worships many gods instead of only one. It is that the idols are the work of man's own hands—they are things, and man bows down and worships things; worships that which he has created himself. In doing so he transforms himself into a thing. He transfers to the things of his creation the attributes of his own life, and instead of experiencing himself as the creating person, he is in touch with himself only by the worship of the idol. He has become estranged from his own life forces, from the wealth of his own potentialities, and is in touch with himself only in the indirect way of submission to life frozen in the idols. (1961, 44)

Historically, idols stood for many things including godlike figures, the state, the church, a person, or possessions. The practice of idolatry, in this sense, is the worship of something into which humanity has put their own creative powers, and to which now they submit, rather than experiencing themselves in their creative act.

Using these notions, we can understand how genetic technologies function. Our embodiment should be an expression of our own creative powers. With enhancement technologies, though, our bodies may be transformed into objects independent of us, the producers. With enhanced capacities, work at ourselves and our individuality may be alienated from us, the laborer, because work has stopped being part of our nature. Instead, our capacities and even genes may be reified as commodities to be purchased just as the proletarian's labor is.

Philosophers within the genetic enhancement debate have at some point idolized these technologies, ascribing powers to them. For example, gene-editing tools, such as CRISPR-Cas9, give us the possibility of enhancing particular traits. The enhanced traits, as such, would provide the means to a better future in theory. Yet, these philosophers seem to have transferred the attributes of an individual onto these enhanced traits, which in turn, strip away our creative powers from ourselves. In this sense, the enhanced traits as objects become idols, and in turn, become the focus of moral debates, rather than conceiving of our lives or the lives of our future children as a whole.

At other points, those in the enhancement debate have been guilty of fetishism, which occurs when one bestows supernatural powers to objects or technologies that those capabilities or technologies in themselves do not possess. Consider Buchannan et al.'s observation:

> [w]hether a given trait will be present depends not just on the gene or genes in question, but also on the environment, including the environment of the organism's body at a particular stage in the organism's development. In the vocabulary of social anthropology, genetic determinism is a variety of fetishism. (2000, 23)

While humans have felt helpless in their lack of control over nature—i.e., suffering, pain and eventually death—this desire to control one's or one's progeny's bodily destiny is still caught up within the dialectical development of human history, especially with the institution of private property. One's body by self-determination is taken to be one's property.

For Marx, when a potter molds wet clay into a pot, that clay retains its use-value because the product, i.e., the pot, is still tied to its material use. When the pot is sold in a store and then purchased by a consumer, the pot loses its value through the monetary transaction. The consumer in turn values the pot, rather than the laborer and the social relation between people becomes lost in the capitalist exchange.

Markus Gabriel expands Marx's conception of fetishism to include one's own identity and lived experience within a society. According to Gabriel, one's

> projection is conducted in order to integrate one's own identity into a rational whole. Understanding ourselves as a part of a rational whole equips us with a feeling of security, of somehow being at ease with how things are, etc. It is easier for us to live with the thought that things (including social institutions) are by themselves rule-governed than with the thought that we must take care to ensure social cooperation so that the social order does not fall apart (2015, 151).

What occurs in fetishism is that we project a structure onto an object, and it is in this sense that our bodies, our identities, even our place within society, become distant from us (ibid., 152). For genetic technologies, my body has now become an object rather than a feature of my subjectivity. In other words, my body becomes an object of social practice rather than an extension of my differentiated subjectivity. It is in this sense that we become both agents and objects. (Meekosha 2010, 1–2)

It is in the reduction to genetically enhanced capacities that we reduce our subjective embodiment, and our human experience, to a *fetishized commodity* in two ways. First, consider the enhancement of artistic appreciation. To train myself to appreciate art, I have to labor at this task. I choose to labor upon this task because I believe that I will find it rewarding and enriching. My accomplishment of being able to appreciate an artwork or piece of music provides me with a feeling of satisfaction, and thus, increases my subjective well-being. If it were the case that my appreciation for art had been instilled since I was an embryo, by contrast, then my choice to labor might be removed. It may be the case that I appreciate art, because that is what I have always done. Or it may be the case that, as a teenager, I rebel against my parents' wishes and eschew all pieces of art.

Second, in the case of well-being, I have transferred the aim of a good quality of life onto a series of skills or objects I must obtain to be happy and find fulfillment. I objectify my body as a set of enhanced qualities and skills, rather than recognize my own differentiated subjective identity. If the goal is to enhance human experience and overall well-being, then we should not confuse the parts (i.e., capacities) for the whole (i.e., a well-lived life) (Gooding et al. 2002).

One's ability to direct one's life is inhibited by attributing powers to false idols we call genetic tools. Regardless of their application which may be jeopardized by human error, the tools themselves do not actually contain the power to cure, the power to heal, and the power to prevent various bodily vulnerabilities. Science is not certitude; technology can only offer probabilities. Moreover, these technologies are not omnipotent. Ascribing powers to them, even if only in theory, does not enable them to accomplish what we hope. Instead, the adoption of these technologies is more likely to impose further the stigma of "abnormal" bodies.

Stigma and Storytelling

At this point, I would like to return to Lorde's discussion of her prosthetic. In her description, we witness stigmatization at the intersection of gender, race, class, and ability. Lorde describes her lived experience as a Black woman being given a pink prosthesis, which looked grotesquely pale under her bra. Here, Lorde emphasizes the difference in the lived experience of illness among women.

Trying to deny the reality of her lived experience as a survivor was not the answer. Instead, Lorde said she was given a choice to either love her one-breasted body or remain alien unto herself. In the space of medicine,

Lorde's body was viewed as "abnormal" and stigmatized, due to the power held by those deemed medical "experts."

Erving Goffman (1963) notes that the word stigma originates from the Greek practice of branding or marking slaves and criminals as well as from the wounds of saints in Christianity (1–2). Stigma, Goffman remarks, concerns "different" physical or behavioral traits and devalues those traits. Thus, to stigmatize someone is to mark that person as different, abnormal, or deviant, and assign negative values to their "difference" (Purcell 2014). With this stigmatization, people with disabilities were cast into negative roles such as "wastes" or "rubbish," "objects of pity," "burdens of charity," or "non-human" (Race, Boxall, & Carson 2005, 510). What can be noted about stigma is its double function: it both designates or brands a person as abnormal and simultaneously devalues that person. What stigma about illness and disability assumes for its operation is a picture of the "healthy" body as normal, which thus renders all other different bodies as "bad."

Consider the case of David Jay's 2011 *The SCAR Project*, which was a photographic exhibition showcasing the portraits of women under the age of forty displaying the scars of mastectomies and reconstructive surgeries (Waples 2014, 164). In 2013, this project became involved in a censorship battle with Facebook. After a Change.org petition, Facebook ruled that "photos with fully exposed breasts, particularly if they're unaffected by surgery, do violate Facebook's Terms," but "post-mastectomy photos are acceptable insofar as they serve to 'raise awareness,' but tread the blurred line of 'graphic content'" (ibid., 165).

In *The Birth of the Clinic* (1973), Michel Foucault emphasizes the way medical knowledge had become a space of power. This power has been used to designate "normal" and "abnormal" as well as labels of sick, unwell, well, ill, able, or disabled (cf. Hall 2017; Hinson and Sword 2019). And so, the power to "cure" the sick or unwell was endowed to the medical practitioner. Those whose bodies were not deemed well or could not be "cured," then, were stigmatized.

The devaluation of different bodies which is at work in Facebook's response to the *SCAR Project* and is present in Marisa Acocella Marchetto's graphic memoir *Cancer Vixen,* reproduces "iconic, heteronormative femininity with commodities and cosmetics" (Waples 2014, 166). We must recognize the "regime of medicalization" which admonishes women to maintain a "normal" and "hetero-feminine" appearance (Bolaki 2019, 111).[11]

Eli Clare talks about the resulting shame from stigma and alienation. He writes, "Shame is a chasm of loathing lodged in our body-minds, a seemingly impenetrable fog, an unspeakable and unspoken fist. It has often become our home" (2017, 163). We become "experts" at hiding our shame; even if we try to tell ourselves differently with affirmations taped "to our bathroom mirrors." (ibid., 166). It won't matter because we won't believe them (ibid.). And Clare cautions us, because it is shame that "hooks us into" the idea of a cure; the way in which it pushes us to pursue "normality" (ibid., 167).

Lorde addresses how stigma and cultural stereotypes consider women with disabilities as unattractive and asexual and denies these labels. Instead, she designed her clothing and jewelry in asymmetrical patterns because women with one breast can be beautiful. In the chapter "Breast Cancer: Power vs. Prosthesis," Lorde specifically considers the intersection of race, class and ability when she talks about the aim of profit guiding treatment rather than prevention guiding the care of black and poor women who have cancer. Here she calls for a "single-breasted community of Amazons" to rise up against capitalism because personal health is political (Bolacki 2019, 112). Instead of seeking happiness and positivity, one should seek the "power and rewards of self-conscious living" (ibid.).

To tell one's story involves the risk of not being heard and the risk of being alienated, misrepresented and othered. To tell one's story is to (re)enact aspects of ourselves and our relations with others. Receiving the diagnosis takes one's voice away, and there is a need to recover that voice rather than remain silenced.

Lorde's work embraces the radically altered body and the radically altered self (ibid., 113). Facing death allows us to live close to the "vulnerable and uncertain *flesh*" (ibid.). Similarly, Arthur Frank in *The Wounded Storyteller* (1995) develops his concept of the "communicative body," which "communes its story with others" and "invites others to recognize themselves in it" (ibid.; cf. Haliburton 2014). Further, Lorde demonstrates a sort of "politicized patienthood" that is centered on "experience" (ibid., 113).

One example has been to turn to the digital space and to find online communities to share one's story. On these platforms, tellers and listeners can confront the stigma and shame by their lived experience. While in the physical space, these community members have been silenced, in the virtual space, they have a voice. In their experience with physicians, they have been ignored, unheard, or dismissed. The listener must not only seek to understand but must also practice belief and openness. On these platforms, tellers have the "right" to speak their "own truth" in their "own words" (Frank 1995). When listeners and tellers create a community, they shape and develop each other. They create meaning in the disruption their diagnosis has created. This experience, I think, can be found in the virtue of relational authenticity, which I learned from Chivel.

THE VIRTUE OF RELATIONAL AUTHENTICITY

Chivel's brain cancer diagnosis shocked our family. Hugo, her husband, is a medical doctor and fought to get her every treatment available. When Chivel's hair began to fall out due to the chemotherapy, her youngest son Hugo Jr. shaved his curly, untamed locks in solidarity and to *speak with* his mother. Erica, the eldest, moved up the date of her wedding and married her fiancé, Mario, so that her mother could be a part of it. As a family, they bore this burden together; they bore their vulnerability in solidarity.

After some months, Chivel's cancer treatments were not working. She lost her vision and her motility. Hugo did not have any more money to pay for her health care. Although their faith gave them strength, they braced themselves for the inevitable future. Still, since Hugo was a doctor, he found out about one last experimental treatment Chivel could try, but because it would be available only at a private institution, it would cost $20,000 (in U.S. dollars). Because in México health insurance works poorly, whatever treatment was not performed at a state institution, where Hugo worked, had to be paid in full by the family.

Jessica, the middle daughter, had always been the good student. Her mother had supported her education and wanted her to become an orthodontist even though it was against the prevailing customs of México. In order to save money for tuition, Jessica worked for years teaching English. In solidarity with her family, Jessica offered to give her tuition money for the chance to save her mother's life. But Chivel refused. Preparing herself for the end of her life, Chivel told her daughter "I will not sacrifice your future. Now my death has a purpose." Jessica pleaded with her, but Chivel would not take the money. She just said "*te amaré siempre*" (I will always love you). Six months later, Chivel passed on.

In her life story, Chivel reclaimed her authorship by choosing her meaning and her purpose. Chivel's life story is not an isolated incident. Rather, it common to us all because we all age and will die some day: it is the realization of our utmost possibilities, which is our vulnerability, our fragility.

Chivel's story of courage and solidarity aligns with Lorde's personal and political stories in her writings. It is these pathographies about facing death that give us a road map and invite us to take part in an exhortation to live well with love for others. As Lorde writes in a journal entry from December 1978, "What is there possibly left for us to be afraid of, after we have dealt face to face with death and not embraced it? Once I accept the existence of dying, as a life process, who can ever have power over me again?" (1988, 110–12).[12]

I think the virtue of relational authenticity aids in giving us the power to enact solidarity with others when facing death. It is this virtue that helps us reclaim the authorship of our life story because our stories are interlocked with the stories of others. The virtue of relational authenticity is inclusive and counters stigma and shame. As a virtue, it gives value to the person herself and gives her life value.

Two Features of Relational Authenticity

As a virtue, to enact one's relational authenticity is to recognize that one's identity and life story are intertwined with co-belonging and sociality (cf. Mackenzie 2014, 2015). In essence, it recognizes that the "who" of who we are is not alone. And it builds self-respect through listening to the personal stories of others. In her essay, "Asymmetrical Reciprocity: On Moral Respect, Wonder and Enlarged Thought," Young notes that subjects "meet across distances of time and space and can touch, share, and overlap their interests" (1997, 351).[13] Moreover, each person "brings to their relationships a history and structured positioning that makes them different from one another, with their own shape, trajectory, and configuration of forces" (ibid.).[14] It is by listening to the personal story of another that I help heal the self.

And by telling my story, those who listen help heal me. It is in the act of telling that I gain self-trust. As a partial standard for authenticity, Anita Silvers and Leslie Francis contend, that authentic ideas and discussions cannot emerge unless the relationship is a trusting one (2010, 254). I have in mind their reconfiguration of the good. For Francis and Silvers, the foundation of this understanding of the good is formed through trust (ibid., 246–47). Self-respect and self-trust, then, are developed through authentic listening and telling with others.

Allison Weir builds upon these ideas arguing that our identity is forged "through both relations of power and relations of mutuality and love. They are intertwined, but neither cancels out the other. Neither can be reduced to the other. Nor do they exist only in radical opposition, or oscillation" (Weir 2008, 12). When children engage with and connect with their parents and others, they open themselves up to knowledge. This knowledge "leads to stronger, more aware, and deeper connections" (ibid.). Through our engagement with others, we find our voice. We do so through the sharing of stories regardless of normative time and space.

Application to Facing Death

Cripping time and space for relational authenticity come together in the life story of Julie Yip-Williams, who described her life as living on borrowed time (Yip-Williams 2018). Born blind in Vietnam, when she was under the age of three years old, her grandmother took her to an herbalist to have her killed (ibid.). Her grandmother thought Julie would never amount to anything, never get married, and never have children. It would have been a mercy killing. But, the herbalist refused.

After that, Julie's parents took her to California, where a surgeon was able to change her eyesight to the status of "legally blind." Later, she earned her way into Harvard law school, got married and had a family. At age thirty-seven she received her diagnosis of stage IV colon cancer. She was given five years to live. All treatments failed and the cancer spread. She then decided to keep a blog so that her daughters and family could have her words. Her blog became a memoir, which was released after her death. She saw death as an opportunity and faced it with gratitude.

Yip-Williams said she started her blog as a way to record her thoughts for her daughters Mia and Belle "if this cancer-fighting journey doesn't end in the way we all hope it ends" (Jordan 2019). This blog became where she could "carve out [her] own little space out there to express [her] sadness, anger, joy, hope despair and a slew of other emotions that come with living with cancer" (ibid.). She searched for books on preparation for death and could find none. She was struck by how American culture lived in denial since all of us die someday. Her hope was to have her book be an exhortation to live, to embrace vividly the hard truths of life, and to be a voice in a silenced genre. Yip-Williams died on March 19, 2018, at the age of forty-two.

Relational Authenticity involves the practice of storytelling both because our identities are intertwined with the stories of others and because we are both tellers and listeners in the authentic communication of our lives. It is the virtue we witness enacted by Julie and her family. Telling stories combines various modalities and voices around a single event or it can be told as a series of events. Storytelling "builds communal bonds," strengthens relationships and preserves or heals the self (Hinson and Sword 2019, 1). It enables us as authors of our own lives to give purpose to the events that befall us.

Julie's life story introduces how crip time affects our authenticity. Crip time draws a distinction between narrative and story. Narrative itself may lack plot and coherence, but it is temporarily linear in its order. Stories, by contrast, need not be told in a linear and temporal order. By drawing the distinction between narrative and story, crip time shifts our perspective for the face of death as a lived experience. It reminds us that we are still the authors of our life stories. Facing death challenges us to bring order to the informal, uncontrollable and chaotic events in our lives by writing and telling a new narrative. Telling one's story is to piece together one's experiences (ibid.). Our narrative can come to be a collection of stories told over time to represent the "self" across various spaces.

Storytelling empowers the teller and enables them creatively to give meaning to their lived experience. Sharing one's lived experience is an authentic relationship, which requires both the teller and the listener to encounter, grapple with and deconstruct the labels they have been given. Labels such as "sick, well, able, disabled, bereaved, or even victorious" (ibid., 1–2). It is an act together of collectively shaping new discourse and altering the perspective of the beliefs, values and categories created by society.

Cripping space in storytelling is to remember that an exchange with another as teller and listener need not be bound to a physical location. As in the case of Julie, her story happened in the digital sphere of blogging. For Young, a condition of communication is that we "acknowledge difference, interval, that others drag behind them shadows and histories, scars and traces, that do not become present in our communication" (1997, 354). To acknowledge this difference requires being open and having moral humility. For example, Young writes:

> I can listen to a person in a wheelchair explain her feelings about her work, or frustrations she has with transportation access. Her descriptions of her life, and the relation of her physical situation to the social possibilities available to her, will point out aspects of her situation that I would not have thought of without her explanation. In this way I come to an understanding of her point of view . . . Understanding the other person's perspective as a result of her expression to me and my inferences from that expression thus continue to carry my humble recognition that I cannot put myself in her position. (1997, 354–355)[15]

The digital space provided the platform for Julie to share with her loved ones and with other listeners her point of view. In her memoir, Julie describes her lived experience of shifting back and forth from optimism to despair. She tries to make bargains with God, she posts pictures of "contented normalcy on Facebook—of meals cooked, a car purchased"—but these pictures are not the reality of the rage she feels toward healthy people, the universe, and other moms she encounters at a birthday party (Gottlieb 2019). Silently, she wants to scream that she did not deserve this fate. In the end, she tries every possible treatment and finally accepts that nothing will keep her alive.

One of the things Julie writes to her daughters is that "paradoxes abound in life." She admonishes them to confront these paradoxes head on. In the face of death, she exudes vibrancy, electricity, and humor along with astute observation saying things like "Health is wasted on the healthy, and life is wasted on the living" (ibid.).

While some assert that "dying is not an option," Julie instead prepares for death and reflects on the life she will one day miss; the life that includes loading and unloading the dishwasher with her husband, making grocery store runs, watching TV with her family, and taking her daughters to school (ibid.). In the end, it is about accepting what we can and cannot control. As she writes, "we can control the effort we have put into living" (ibid.). It is up to us to see the beauty and meaning in the moment, in our daily lives, and in our relationships.

Relational authenticity, then, gives us our voice back when the face of death has silenced us and tried to strip away the authorship of our life story. It does this through empowering us to tell our story with others. This means to conceive of our life and the lives of others as authentic wholes. As a virtue, it enables our voice and helps us heal when we have been silenced. Moreover, relational authenticity is enacted with others, and as such, empowers us to reclaim the authorship of our life story. It helps us answer the question that Lorde asks of us: how do we channel our power into the service of what we believe?

CONCLUDING THOUGHTS

My lived experience with Tía Chivel taught me the answer to this question. The story of Chivel and her family taught me the value of relational authenticity when we come to terms with our own mortality. Her courage and her family's courage, enacted in solidarity, show that our relationships with others are intertwined in our personal stories. Our storytelling gives both the teller and the listener the ability to act together to collectively shape human experience.

While the promise of genetic enhancements is to give us borrowed time, i.e., additional time, the reality is that they will not be able to alleviate our fear of mortality as a face of vulnerability. They promise a future we can control. Yet, we are ascribing powers to tools we have created. Genetic technologies can neither rid us of nor postpone an inevitable feature of the human condition.

In the next chapter, I return to one of my own stories of vulnerability and address the face of misfortune. The face of misfortune has mistakenly been linked to living with a disability, a connection which many philosophers have associated with race and class as well. I show instead that the lived experience of misfortune as loss or deprivation many times is due to a confusion of our needs with imaginary appetites. Instead, by embracing the virtue of empowered self-direction can we then view the face of misfortune with a better perspective.

NOTES

1. Jeff Mason was a lecturer with Middlesex University. I follow his story of living with terminal lung cancer in (Mason 2011) and (Mason 2012).

2. Vanessa's story can be found on YouTube, "My Walk to Remember (Diagnosed with Colorectal Cancer)" at the following link: https://youtu.be/WcbMltqvQkE

3. Tamara O'Brien passed away on October 15, 2019. Her story can be found on YouTube at the following link https://youtu.be/czlMz5IhUww

4. For example, the Look Good Feel Better program https://lookgoodfeelbetter.org/

5. In her foreword to the 1985 British edition of The Cancer Journals, naturopath Carol Smith praised Lorde's "strengths and insights," arguing that "prosthesis is a part of the spectrum of oppression that treats people with physical differences or disabilities as 'abnormal.'"

6. Etieyibo (2012) critiques Alhoff's proposal for ignoring the importance of non-genetic factors in the development of our capacities.

7. Names have been changed for the purpose of publication.

8. In the sense of facing death rather than being wrongfully killed.

9. In *Raising a Rare Girl,* Heather Lanier tells the story of meeting a woman in a park who shares the personal story of her son "beating" cancer and asks Lanier about a "cure" for her daughter. The idea of a "cure" is what many facing death hope for and pray for. It is in fixing the "problem" that one finds one's victory (2020, 272–73).

10. According to Fromm (1961), the concept is when one's existence is alienated from one's essence.

11. One of the issues can concern hypervisibility through commercialization and "pinkwashing" of the "cancer industrial complex," which is "the multinational corporate enterprise that with the one hand, doles out carcinogens and disease, and with the other, offers expensive, semi-toxic pharmaceutical treatments" (Waples 2014, 163).

12. See also Tracy Smith (2020), "How Audre Lorde's Experience of Breast Cancer Fortified Her Revolutionary Politics" at the following link: https://lithub.com/how-audre-lordes-experience-of-breast-cancer-fortified-her-revolutionary-politics/

13. Young (1997) offers two perspectives to take into account for relational authenticity: temporality and specificity of position. For temporality, each subject has her own history and brings this history into communication with "particular experiences, assumptions, meanings, symbolic associations, and so on" (352). Temporality, however, is incomplete without specificity of position. This history intersects with one's social position, which can flow and shift, but will be contextualized within relations (ibid., 354). Cripping time and space, then, help enact relational authenticity.

14. Margueritte La Caze (2008) argues that friendship and political representation complicate Young's treatment of self-respect for asymmetrical reciprocity.

15. Our relatedness to others plays an important role in our self-conceptions and the dynamics of deliberation and reasoning. Our self-conceptions are fostered by care, which includes both helping someone and giving warmth and affection; one must not confuse the two (Asch 1993, 118).

Chapter 4

Facing Misfortune

Perfection, Designer Babies, and Imaginary Appetites

INTRODUCTION

This chapter weaves together the experience of misfortune as a face of vulnerability and the tricks our imagination can play on us. In adoption counseling, the discussion turns on two kinds of loss that those involved experience. There is a loss that the birth mother experiences, because she has had to make the difficult decision to give up her biological child. There is also the loss the adoptive parent(s) experience. Many times, it is the loss of infertility. Both sides experience misfortune and grieve. Yet, it is the love for the future child which makes these losses something both sides can work through and overcome.

As the adoption experience has taught me, it is neither hope nor acceptance that helps us encounter the face of vulnerability that is misfortune. Experiencing misfortune in life is a part of the human condition. It is entangled in the vulnerability of human living. To have hope means that we open ourselves up to the possibility of despair. To have dreams and goals about our future means that many of those will not become reality. And to anticipate future projects means that some of them will fail.

Genetic technologies cannot erase the face of vulnerability that is misfortune. And yet many proponents assume that they will be able to do so. My suspicion is that these technologies cater to what Marx would call imaginary appetites, which were created by industry. In the pursuance of these imaginary appetites for ourselves or our future children, we run the risk of commodifying human experience, thereby losing all sense of its intrinsic value.

In this chapter, I address the face of misfortune, why we fear it, and how the capacities promised from genetic technologies do not shield us from human vulnerability. The stories which I draw upon include my own struggle with infertility and the story Heather Lanier tells in *Raising a Rare Girl.* I weave these stories together in order to reveal the limit genetic technologies have in their promise for a certain future. This promise, I argue, is due to a fault in our imagination. In contrast, I show how encountering the face of misfortune requires self-awareness and a kind of self-direction which is empowered because the sense of who we are and who we will become is interdependent with others. Perhaps it is best to begin with my experience facing misfortune—the experience of infertility—to explain this face of vulnerability.

MISFORTUNE AS A FACE OF VULNERABILITY

I hung up the phone after I got the news. My pregnancy test was negative. This disappointment had become a monthly routine. Trying to conceive by following the advice of the fertility clinic, my husband and I embraced the ketogenic diet, slathered coconut oil on every meal, participated in daily yoga, meditation, and stress-reducing activities along with consuming every supplement they recommended. The hormonal treatments turned me into an insecure, confused teenager trying to maintain composure and efficacy in an adult world. Each week I would make my one-hour drive to the clinic to receive treatments or get blood tests. Each week we would wait patiently, hoping, praying, pleading for a test to be positive. Just one.

That day never came. The pain and confusion which come with grief is difficult to describe, but it came out in moments of silent and hidden despair. I couldn't turn on social media on Mothers' Day, even though I felt joy for all of my friends who were mothers. I cried quietly alone in my office or in the bathroom so no one could see my pain. Teaching the lecture on selective reproductive technologies to my students in bioethics became a true test of maintaining equanimity while trying not to have a breakdown. I kept asking what was wrong with me? Why am I broken? And what do I keep doing wrong? We were deemed "healthy" and doing everything right, and yet why were we not pregnant? Medical doctors and a battery of tests gave no answer to this mystery.

Having a child for many people is part of a life plan. It is part of the dream that many of us include in our future projections. Infertility ruptures a person's life plan. My husband and I each grieved differently for a dream that was lost. I felt helpless and empty trying to reconceive of what we could do. The experience, and the eventual acceptance of this loss, altered my self-direction. The loss experienced in infertility is a loss of a part of yourself—the part of your future that will no longer be.

Infertility is one way I have encountered misfortune as a face of vulnerability. Nine observations prove helpful for the present reflection. First, misfortune is not due to something we did, and yet, we still demand to know a cause for why it occurred. We search for something or someone to blame. We struggle to accept that we have been caught in "bad luck."

The face of misfortune, then, reveals itself as events we encounter, even though we search for its cause in actions and bad habits. Misfortune reveals that one's choices in life are subject to chance and that any belief otherwise is a lie. The reason we fear misfortune is because it casts doubt on our quest for certainty.

Our lived experience is taken from a personal stance. Third, this means that our orientation to the world is grounded in our beliefs about how the world is supposed to go: that there are answers to our questions, that there are justifications for our beliefs, and that there are solutions to our problems (cf. Haliburton 2014, 247). We fear misfortune because it reveals the fragility of our future. It calls into question our sense of meaning at the center of our being-in-the world.

The lived experience of misfortune is attuned to the future in normative time. As a fourth observation, the experience of misfortune suggests an alteration in one's self-direction. It entails a stripping away of a part of one's future. Expressed through grief, it is when we have to let go of what we thought would eventually be. We fear misfortune because of whom we will not become or that part of our life plan will never materialize. Misfortune cuts off our future insofar as the *some day* we hoped for, strived for or imagined will not come. It expresses loss in normative time as a loss of what could have been. It is the realization that one's potential will not be actualized. It is the lived experience of failing to be, to do, or to have. It is a falling short of achieving what has been viewed as a good.

Sometimes, as I have observed, misfortune shows its face as a deprivation rather than as a loss of potential. In this fifth sense, misfortune is when our prospects are lower than others. From the perspective of evaluation, it may be that our state is viewed as disadvantageous or that we are badly off. In either sense—as a loss of potential or as a deprivation of some good—misfortune redirects the trajectory of our life plan.

In response, we grieve the loss of what could have been. I think that May interprets grief from a helpful perspective. He says "grief seems a proper recognition of a loss that has taken place both for me and for the person who was there. It is a melancholy for the meaning that has been leaked away from me and the respect for the one who is no longer with me" (May 2017, 126). For misfortune, we grieve because some good we held close to our hearts is now gone.

As a sixth point of observation, in normative space, the lived experience of misfortune can show itself as an empty room, a road no longer taken, or a locked door. We lose our orientation because those places which were once familiar now feel alien to us. Even in our own home, our arrangements and habits seem out of place and strained. Trinkets, which remind us of dreams once dear, are now discarded or disowned. In space, misfortune reveals to us what our concerns with daily minutiae concealed: that our homes are not just the arrangement of objects I call mine, but rather a "sedimented history which is the story of my life, and the story of our shared lives" (Young 2005, 158). Misfortune erodes this sediment of our lives because those things are no longer meaningful in the same way.

Yet, misfortune affects our social lives as a seventh point of observation. When our misfortune becomes known to others, we feel our subjectivity slip away under their gaze. Our family, our friends, and even strangers, express sorrow over our loss of flourishing. In their gaze, we detect that our circumstances have been deemed unfortunate. We detect their gaze and experience ourselves being observed by the other. Indirectly, we become objects under the well-meaning gaze of pity (Shapiro 1994, 3–4).

In misfortune, though, it is more than just observation. It is more than mere judgment that we fear. Through their gaze, I now see myself through their eyes, and become other to myself. Their sorrow for our misfortune strips away our subjectivity. In their gaze, we feel ourselves become othered within their narrative. With sympathy or compassion, they express hope that we will overcome our circumstances. Trapped in another's narrative, we are caught between overcoming our misfortune or succumbing to it. As an object now entangled in the story of another, we must become a hero and overcome our misfortune. If we do not, and instead we succumb to our situation, then we are looked upon with pity, aversion, and even disgust. And in response, shame may become our new home (Clare 2017, 163).[1]

We fear misfortune in part, then, because our subjectivity is stripped away by the sorrow of another. In this eighth sense, we fear misfortune because we become an object of their pity. We will do anything to control our circumstances, avoid our failures, be subject to the whims of the universe in order to retain what is familiar to us: our *selves*.

As a final note, misfortune dictates which stories we share and which stories we choose not to. As Eli Clare notes, "we are experts at hiding our shame, swallowing it, pretending it's not there" (2017, 166). If we had the chance, most of us would choose to prevent misfortune from ever entering our lives. Most of us would choose a certain future we could control.

Genetic technologies give us this promise of a certain future. A future in which particular kinds of misfortune cannot enter. These technologies would have us believe we can retain control in our quest for certainty. They would have us believe that our life plan is not up to chance, but rather, up to choice. The forgoing analysis allows us to make existential sense of that motivation, but let us turn now to an analysis of its delivery.

DESIGNER BABIES AND CONSUMER EUGENICS

In Heather Lanier's *Raising a Rare Girl,* Lanier tells the story of raising her daughter Fiona. Her daughter was born with the condition Wolf Hirschhorn syndrome. When Lanier first learned of her daughter's syndrome, a mother told her that she would always grieve the child she didn't have. Lanier puts it this way: "it was both a warning and a membership card" to the Country Club of eternal grief (2020, 267).

When Lanier was pregnant, she—just like me—followed the doctors' orders. As she says, she tried to make a "SuperBaby": swallowing capsules of mercury-free DHA, eating organic fruits and veggies, avoiding soft cheeses, cold cuts and kept her flip phone at arm's length (ibid., 3). She read every book and tried to do everything right in order to have the "perfect" child.

Lanier points out that her attempt to make a SuperBaby was an act of privilege: she had "privilege in spades, not only from my race, class, and ability, but from my willing co-parent" (ibid., 28). In her memoir, she mentions where this desire to have a "perfect" child seems to come from. It seemed to come from the "culture of motherhood." And this "pressure to make perfect children also appears to have global reach" (ibid., 29). For those mothers who had children with disabilities, though, while they had "done everything right," there was a suspicion of how it might be their fault. This, Lanier concludes, was the messaging in "pregnancy culture" and the "culture of motherhood": you are responsible for you baby's wellness through your choices, both physical and emotional (ibid.).

But when Fiona was born, Lanier's faith in the culture of motherhood was thrown into question. Lanier describes what she had imagined for the episcopal ordination ceremony for her husband: "As Justin bows to a bishop and makes his lifelong vows wearing fancy vestments, I sway back and forth while wearing a peacefully soothed baby in a cotton wrap . . ." (ibid., 34). This image of motherhood, however, was not going to happen. As Lanier continues, "at under five pounds, Fiona didn't qualify to fit inside the great symbol of natural mothering, the cotton wrap, a fact that felt like a metaphor in itself" (ibid.). Instead, her doctor advised her not to go because her "baby was too small for the public" (ibid., 35).

As time passed, Lanier realized that she did not yearn for a different child. Rather she wished that her daughter's body would be viewed as a variation rather than as an anomaly. The world was changing before her very eyes:

rather than accepting human variation, scientists were "making gene edited CRISPR babies" and the first target was a gene for deafness (ibid., 267).

A Custom-Made Child

As Lanier has pointed out, the hype surrounding the idea of "designer babies" is still present in contemporary forums in the twenty-first century (Conrad 2002; Scott 2006; Pavone and Arias 2012). After the completion of the Human Genome Project in 2003, publicity and various misleading promises circulated in the media that the major genes for heart disease, mental health disorders and diabetes would soon be discovered and that genetic cures would be available (Alper et al. 2002). Some critics responded by saying these claims were misleading and harmful notions of genetic essentialism (Nelkin and Lindee 2004): they were misleading because they oversimplified the relation of genes to traits and they were harmful because public resources had been used for research that might have been more fruitful for other public health needs.

The idea behind CRISPR-babies or "designer babies" that haunts the philosophical debate concerns an appeal to reproductive liberty. Discussion sometimes includes the idea of a custom-made child, which views the embryo and fetus as a biological property of the would-be parent(s) (Hughes 1996, 7). The argument goes that just as women are allowed the "reproductive right" or "choice" to select the characteristics of the father of their child, so too future parents should be allowed the right to choose the characteristics of their "custom-made child" from a catalog (Resnik and Vorhaus 2006, 16). Even if it were the case that vast inequalities emerged between the designed babies and the un-designed babies, some philosophers have argued that the individual rights of the parent(s) should take priority (cf. Flanigan 2017).

This debate concerning a "custom-made child" and the potential selective reproductive technologies involved has been referred to as "consumer eugenics" (Gupta 2010). These fertility services have now become commercialized, and as a result, philosophers as well as potential parents, have raised some concerns. In essence, two of these concerns involve viewing the potential child as a product.

The first concern is a practical one and targets the challenges of global policies concerning these technologies. As we have already witnessed with the conflicting global policies on surrogate motherhood as well as the exploitation of women in the Global South by potential parents in the Global North, there have been difficulties implementing any standard guidelines. For example, three different studies concerning policies were conducted in India, Cuba, and Brazil which highlight the decentralized commercialization problem of consumer eugenics. In India, due to the high cost, genetic testing and screening programs are "left largely to the private health sector to offer the services" (Gupta 2010, 46). In contrast, for Cuba, genetic testing and screening programs are centralized and are offered free of charge to the public by the government (Heredero-Baute 2004, 131). For Brazil, however, although the genetic testing and screening programs are offered by the government, the ministry of health has been decentralized. The decentralization has led to the fragmentation of data collection and a loss of comprehensive information available about the proportion of state-financed hospital births covered by newborn screening (Horovitz, de Mattos and Llerena Jr. 2004, 113).

This lack of uniform global policy leads to the second concern, which is of philosophical interest, concerning consumer eugenics: the commodification of a custom-made child (cf. Dhanda 2002). Michael Sandel (2007), for example, has argued that genetically enhancing children treats them more as commodities than as gifts. Since these technologies already operate in a global marketplace, and enhancement technologies will likely be produced and distributed in the same marketplace, it is important to consider the possible commodification of children due to a lack of regulation. Sandel's concern is one shared by many in the public. In one study in the United Kingdom, S.E. Kelly and H.R. Farrimond observed that potential parents raised concerns about the commercial misuse of these technologies (2012, 77).

Many feminist and disability scholars have pointed out that we already commercialize cosmetic procedures in this way (Bordo 1993). The medicalization of human beauty and aesthetic values can influence scientific aims and empirical judgments about what is best for the child and can influence parental aims (Silvers 1998). Without any kind of regulation to keep the economic aims of capital in check, the idea of a custom-made child should give one pause. Commodification, by its nature, replaces the individual worth of something with its exchange value. We tolerate these exchanges with clothes and cars but should not do so with humans. This set of reflections leads to a related point about who could afford these technologies.

Genetic Privilege

Lanier's reflection on her own privilege opens up another issue surrounding genetic enhancements: could germ-line enhancements widen social inequality even further? In other words, could the making of these customized children lead to what might be called "genetic privilege"?

Some philosophers seem to think that it could. Peter Singer (2008), for example, has cautioned that without regulation of these commercialized technologies within the global marketplace, these enhancements would be enjoyed by the wealthy and pass advantageous traits to their descendants. As a result, the gap between the rich and the poor would be further widened.

Likewise, Nicholas Agar (2010) has raised the issue of how these technologies, such as enhanced lifespan (discussed in chapter 2), would be distributed. Would enhancements such as these all go to the wealthy, allowing them to live a thousand extra healthy, young years (2010, 130)?

Genetic privilege, like the other forms of privilege Lanier discussed, would give advantages to the few (Selgelid 2014; Mehlman 2003; Farrelly 2005). Currently, those with wealth already enjoy privileges such as access to more resources, better schools, and larger social networks. In the United States, for example, there is already an imbalance in genetic testing due to intergenerational poverty (Penchaszadeh and Puñales-Morejón 1998, 139). What if the gap for genetic privilege were to become wider? This definitely seems to be a possibility, and one that Bostrom himself entertains. For example, he speculates on what "the haves and the have-nots" might be like in the following way:

> The genetically privileged might become ageless, healthy, super-geniuses of flawless physical beauty, who are graced with a sparkling wit and a disarmingly self-deprecating sense of humor, radiating warmth, empathetic charm, and relaxed confidence. The non-privileged would remain as people are today but perhaps deprived of some their self-respect and suffering occasional bouts of envy. The mobility between the lower and the upper classes might disappear, and a child born to poor parents, lacking genetic enhancements, might find it impossible to successfully compete against the super-children of the rich. Even if no discrimination or exploitation of the lower class occurred, there is still something disturbing about the prospect of a society with such extreme inequalities. (2003, 503)

It could be the case that those with genetic privilege would consider themselves wholly responsible for their success rather than acknowledging the origin of their privilege (Sandel 2007, 92). While Sparrow (2015) has brought up the idea of a future "enhanced rat race," it could become even more complex. With our continual development of these technologies, each group of enhanced individuals could find themselves falling more and more behind has newer enhanced individuals had the advantages for important social and material goods.

While many philosophers worry about a possible dystopian outcome, what I find of particular interest is the approach transhumanists take on preventing genetic privilege. In general, they uphold a welfarist response by addressing how to correct for the lack of policy, regulation, and opportunity using these technologies. For example, Bostrom has argued that one way of responding to our new genetic technologies is to direct our own human evolution (Bostrom 2004). He suggests that social structures could be arranged in such a way to reduce the fitness of "non-eudaemonic types" and instead enhance the fitness of "eudaemonic types" (ibid., 13). For Bostrom, transhumanism "does not entail technological optimism" (Bostrom 2005c, 3). Social structures need to be put in place to avoid the worry that Sparrow and Sandel have and to ensure fair access.

To do so, Bostrom proposes constructing social policies to counteract the inequality-increasing tendencies. He gives the example of widening access for poor families through subsidies in order to benefit everyone. He does acknowledge that not all parents might take advantage of enhancements, and this would result in diminished opportunities for their children by no fault of their own.

But would it be appropriate for the government to intervene? Whether by limiting the reproductive freedom of the parents who desire to use the technologies or by requiring all children to have certain enhancements? Bostrom thinks the latter may be more promising: "By requiring genetic enhancements for everybody to the same degree, we would not only prevent an increase in inequalities but also reap the intrinsic benefits and the positive externalities that would come from the universal application of enhancement technology" (2003, 503). Simply stated, Bostrom and other transhumanists seem to think that requiring particular enhancements, whether for health or morality, is the best way to go forward.

Bostrom does not think that possible unjust inequalities due to technology is a sufficient reason to curtail its development and use. Instead, we should consider the benefits and the intrinsic values it may bring such as good health, intelligence, and emotional well-being. It could be the case, however, that the use of genetic technologies does not result in an increase in social inequality. Instead, it might result in just faster diagnoses and an increased ability to cure genetic defects by eliminating conditions such as Tay Sachs, Lesch-Nyhan, Down syndrome, and early-onset Alzheimer's disease (ibid., 504). According to Bostrom, "[t]his would have a major leveling effect on inequalities, not primarily in the monetary sense, but with respect to the even more fundamental parameters of basic opportunities and quality of life" (ibid.).

I find the transhumanist approach problematic for two reasons. First, it is latent with ableist prejudice. And second, what Bostrom and other transhumanists have suggested, i.e., the requirement of certain enhancements, is technically a form of positive eugenics. It is a form of positive eugenics because it would restrict reproductive liberty. To understand why their proposal is so troubling, some context is necessary.

The Origins of Consumer Eugenics

To understand why there should be concerns regarding the transhumanist proposal, some brief history is needed. In her memoir, Lanier discusses the origin of the Bell curve, known as normal distribution. When the Bell curve had originally been applied to humans, its focus was on the average man, for the nineteenth century, in which normal was good while any deviation was considered bad. Eugenics, however, changed this focus (2020, 150–51; cf. Hernstein and Murray 1996).

Eugenics, a word coined by Francis Galton (cousin of Charles Darwin) in the early 20th century, was initially embraced as a practice. It was argued that humans could apply the knowledge of animal breeders to people in order to "improve the human stock." Galton changed the interpretation of normal distribution by emphasizing that some standard deviations were better than others. For example, tallness and high intelligence became desirable. Galton's legacy is the problem for disability in the debate concerning genetic technologies: the "problem is the way that normalcy is constructed to create the problem of the disabled person" (ibid., 151).

The eugenics program consisted of two components: a negative component, which concerned people with undesirable traits who would not be permitted to reproduce, and a positive component, which concerned people with positive traits who would be encouraged to breed with each other. Eugenicists erroneously believed that single genes were associated with single traits, such as criminality and intelligence. The aim of eugenics, then, was to improve the transmission of desirable traits while limiting the transmission of undesirable traits within a population.

What transhumanists are proposing, though, and what seems to be taking place in the practice of consumer eugenics already, was adopted by many countries when eugenics became a movement. In 1900, many countries, including the United States and Germany, began to implement eugenics programs (Paul 1995). Eugenics included many movements and eugenicists varied considerably in both their methods and their goals. In the United Kingdom, for example, advocates urged voluntary changes in mating patterns (Selgelid 2014, 3–4). The movements in France and Brazil differed in focus insofar as they were concerned with healthy deliveries and child-rearing. In the United States, by contrast, "prizes were handed out at state fairs for 'Fittest Family,' alongside the prizes for best cow and best pig, to very ordinary farming people" in states such as Kansas and Arkansas (Wikler 2009, 343). Even the clergy were encouraged to use eugenic guidance in the counseling for couples considering reproduction (Crook 2008).

After World War II, eugenics fell out of favor. Part of this was due to the Nazi atrocities. Social commentators argued that "fitness" and "unfitness" were racist, sexist, and classist and were not used in science. In the 1940s and 1950s, social sciences such as sociology and anthropology began to link criminality, destitution, and mental illness to social and economic forces, rather than genes. In science, it was demonstrated that traits such as intelligence and antisocial behavior were not the result of any single gene but were the result of the interaction of multiple genetic and environmental factors throughout childhood and adolescent development (Conrad 1997). These scientific and social scientific findings made clear that the tactics of negative eugenics would not eliminate undesirable traits (Geller 2002).

By the 1950s, eugenics programs, journals and organizations rebranded themselves as "medical genetics," "heredity clinics," "social biology," and "human genetics" even though the aim remained the same: use the insights of genetics to control human heredity. Using similar techniques, reproductive counseling clinics emerged. In these clinics, geneticists asked couples questions about their medical histories to determine which couples were at risk of bearing a child with a disability (Wertz 1998). Those who were at risk were advised not to procreate (Iltis 2016). In the 1950s to the 1960s, the services these clinics could offer expanded with the molecular revolution in biology: carrier screening as well as amniocentesis were now available.

There was a tension, however, at work in the community of genetic counselors. On the one hand, the tools used such as a family pedigree and screening for traits such as Down syndrome had been used by heredity clinics and eugenicists before. On the other hand, the genetic counselors were mostly women who had been influenced by the criticism from second wave feminism, disability rights and bioethics (Ajf-b 1983). To alleviate this tension, genetic counselors utilized the principle of nondirectiveness: rather than tell patients how to act, they would let patients decide for themselves how to act on the genetic information they received (Buchanan et al. 2000). In short, the counselors would respect the reproductive liberty of the potential parents. As a result, from then on throughout the 1980s and 1990s, researchers had available an ever-expanding pool of medical genetics to find the precise genomic location of disease-causing genes.

Disability scholars, however, have argued that the genetic counseling centers were many times biased against children with disabilities (Parens and Asch 1999; Carlson 2001; Hall 2013). Stated differently, while the genetic counselors claimed that their non-directive approach was unbiased, especially in the case of disabilities, they merely changed their rhetoric (Alper *et al.* 2002; Wendell 1996). The intersection among health disparity, race and disability was also present insofar as medical resources were pulled away from treating racial health disparities and re-allocated to genetic technologies (Roberts 2009).

The history of the eugenics movement, thus, led to the widespread support to ensure reproductive liberty for individuals and couples regarding their decisions (Beckwith 2002). Couples and individuals in the United States are now free to decide if they wish to have children and to determine when and how many children they have, and even free to decide which reasons to guide their decisions (including no reason at all) (Wachbroit and Wasserman 2003).

Some of the biases of the positive eugenics programs, however, are still present in the contemporary debate concerning genetic enhancement. While most no longer hold to "genetic essentialism" in theory, these biases are often present in practice and become apparent in contemporary genetic counseling (Dar-Nimrod and Heine 2011; cf. Wasserman and Asch 2012a and 2012b).[2] The transhumanist proposal of requiring certain enhancements, then, would be considered a form of positive eugenics, and should make us hesitate moving forward. And yet, the question remains, why would we want to have the custom-made children transhumanists think we do?

IMAGINARY APPETITES AND COLONIZING MISFORTUNE

As Lanier herself asks, why did she want a SuperBaby? And why is the idea of the perfect child so intertwined with the culture of motherhood? I do not think that the answer to these question lies in the history of the eugenics program alone. Rather, I think that the answer to these questions lies, in part, in Marx's concept of imaginary appetites.

Imaginary Appetites, the Focusing Illusion and Misfortune

Marx draws a distinction between human drives and appetites. Human drives are constant and include drives such as hunger and sexual urges, which are part of human nature. Appetites, by contrast, are not part of human nature. Instead, they arise from social structures in certain conditions of production and communication (Fromm 1961, 25). Capitalism incentivizes the confusion of our actual needs with imaginary appetites for profit.

Every new product is now transformed into a need to satisfy fantasy, caprice and fancy (Fromm 1961, 55). The entrepreneur awakens in us unhealthy appetites, and as Fromm notes, following Marx, the "man who has thus become subject to his alienated needs" has become a "self-acting commodity," who only knows how to relate to the world by consuming it or using it (Fromm 1961, 56). The aim of the production of capital is to place humans in the role of commodities. This is the type of consumer and laborer capitalism needs; it needs the passive recipient that consumes things to satisfy synthetic needs (Fromm 1961, 57).

When Lanier described how she imagined herself at Justin's ordination ceremony, i.e., embodying the ideal of motherhood while wearing a peacefully soothed Superbaby in a cotton wrap, in contrast to the reality of staying home, because Fiona was too small, we can recognize that the image Lanier had in mind was given to her by capitalism. That is, capitalism has created the imaginary appetite of the SuperBaby or designer custom-made child. The designer baby as an idea was created by capitalism, and we have confused our desire to have a family with a concept created by the intersection of science and capital.

Similarly, the culture of motherhood with the idea of the "normal" mother has been promoted by capitalist incentives. What do we gain by having the "perfect" child? What do we gain by being praised for being the ideal mother? We gain recognition, which leads only to another kind of rat race. Recognition for accomplishments gives a sense of reward and satisfaction that quickly fades. After it fades, we end up wanting more, or wanting to do something again to bring back those feelings. As Todd May notes, this feeling lets us know that our sense of who we are and our self-worth come from the outside (May 2017, 108). It is the desire for "something to be granted to me in the future that will make me feel good or worthy or whole or sufficient" (May 2017, 108). Consumerism includes this desire for recognition. It creates a "false self, a self layered over the true self, one that is eternally anxious and unsatisfied" (May 2017, 108).

This desire for an imaginary appetite, and the recognition we imagine we will receive for attaining it, might be attributed to what psychologists call *the focusing illusion.* The focusing illusion is an error in imagination. It occurs when "people overestimate the emotional impact of events by disproportionately focusing on narrow life domains influenced by the events" (Ubel, Loewenstein, and Jepson 2003, 601).

The focusing illusion is best illustrated by twin results of lottery winners and those with paraplegia. Researchers found that despite what most suppose true, the "happiness of people who had recently developed paraplegia or quadriplegia following a motor vehicle accident did not differ substantially from that of recent lottery winners" (ibid., 599; cf. Hope 2011). What is at work is a predictable error in imagination. The error in imagining paraplegia is that one's attention is drawn to the loss of activities due to paraplegia, such as playing favorite sports or dancing. As a result, one ignores other activities unaffected by paraplegia, such as spending time with one's family or watching favorite television shows (ibid., 601).

Let me provide another example of the focusing illusion, which highlights how the error occurs when our attention is drawn to particular details. Ubel et al. found the following in their study:

> For example, in one study, college students in California and Michigan were asked to state how happy they are currently, and how happy they would be living in the alternate state (e.g. how happy do California students think they

> would be living in Michigan). Both groups of students reported similar levels of current happiness. Nevertheless, both groups also predicted that they would be happier living in California than in Michigan. The size of this predicted difference was correlated with students' beliefs about the relative impact that weather has on their happiness. This study has been interpreted as evidence that students focused too narrowly on weather when making these predictions, ignoring the possibility that California might have more traffic jams, a higher cost of living, or other features that on balance make life just as good in Michigan. (ibid., 601–02)

Just like the students, we fail to consider the big picture because we cannot fill in all the details. The point is that the focusing illusion may be at play in the desire to have a SuperBaby and the desire to be an ideal mother. Both are images attached to products to capture our imagination. Rather than considering all the different aspects of being a parent—changing diapers, calming temper tantrums, sleepless nights, and moments of great joy—the images which come to mind are those given to us by capitalism.

And when we do not attain these imaginary appetites—what has been the object of our focus for so long—we feel the lived experience of misfortune from failing to be, to do or to have what we thought was a good. Similarly, if we feel that we have been deprived of something, we have to ask ourselves what kind of evaluation are we using? And what kind of bias might be present?

In her memoir, Lanier catches herself feeling this loss for Fiona's intelligence, because intelligence was something she had always valued. In response to Lanier's expression of grief, Justin replies, "she's not damaged goods, you know" (2020, 126–27). In that moment, Lanier recognizes her own bias. As she struggles to reconcile two opposing beliefs, she asks Justin, "What will she do with her life? . . . What kind of life will she lead?" He answers: "She'll live her life" (ibid., 127).

The association of disability with loss and misfortune is due in part to the focusing illusion. The focusing illusion explains the "tendency that causes us to overestimate the happiness of Californians also causes us to underestimate the happiness of people with chronic illnesses or disabilities" (Gilbert 2006, 114). The focusing illusion also explains the imaginary appetites we pursue such as the ideas of the perfect child or the ideal mother.

Could it be possible that the focusing illusion is present in the genetic enhancement debate as well? I believe so. Consider Nick Bostrom's "Letter from Utopia" for example. Bostrom imagines what life might be like in the future after humans have been enhanced:

> Does the whole exceed the sum of the parts or do the parts exceed the whole? What I have is not more of what you have. It's not only the particular things, the

> paintings and toothpaste-tube designs, the book covers, the epochs, the loves, the rusted leaves, the rivers, and the random encounters, the satellite photos, and the hadron collider data streams. It is also the complex relationships between these particulars. There are ideas that can be formed only on top of such a wide experience base, and there are depths that can only be plumbed with such ideas. And the games. And the lusty things, and the things I can't even mention. (2008a, 3)

In an earlier essay, "Transhumanist Values," Bostrom imagines what life today could have been like if this technology had been available before the present time. His reflections include the possibility that earlier humans would have enjoyed health, youthful vitality and even reached "levels of maturity that we can barely imagine" (2005c, 6; cf. Bostrom and Sandberg 2008).

Bostrom's beliefs about the future, however, are likely subject to the focusing illusion. We must remember that just "as we pass along our genes in an effort to create people whose faces look like ours, so too do we pass along our beliefs in an effort to create people whose minds think like ours" (Gilbert 2006, 236). Transhumanism holds to certain beliefs about enhancement technologies and eliminating particular traits from embryos. It may be that some of the beliefs they hold follow from too narrow a scope of awareness.

Yet, I do not believe that imaginary appetites and the focusing illusion can explain the whole story. The association of disability with the face of misfortune, including the desire to eliminate certain traits in embryos, has another origin from the history of capital: the history of colonialization.

Colonization, Misfortune, and Disability

In my research in philosophy on the topic of misfortune, I discovered that philosophers did not conceive of misfortune quite like I did. Often, they fell into discussions of comparison for understanding misfortune in contrast to good fortune in life. Moreover, they wanted to know if the events involved in misfortune were partially bad or just plain bad overall. It is because we assume a standard of evaluation to determine who is flourishing and who is unfortunate. And in their consideration of misfortune from a comparative framework, many of them thought that living with a disability was unfortunate and made life worse.

One of the philosophers who holds this perspective is Jeff McMahan. In his article, "Cognitive Disability, Misfortune, and Justice" (1996), McMahan uses offensive characterizations of people with disabilities as examples of misfortune. In various ways, he argues that those with impairments are deprived of different dimensions of the good life or have lost the potential they once had for greatness (pp. 8–17). The link between disability and the face of misfortune, especially within bioethics, though, has a long history, so McMahan's ableism is nothing new (Tremain 2017, 19).

There seem to be three overlapping biased beliefs that lead to the association of living with a disability to being unfortunate (Rohwerder 2018, 5–8). As I mentioned in section two, when we experience misfortune, we often search for a cause or an explanation for why things have happened to us. Present in two of the beliefs which associate disability with misfortune, there is a bias which holds that having a disability is a kind of punishment for a bad deed, an act of God's will, or the result of some kind of magic or witchcraft. As a result of these beliefs, many people with disabilities have been subjected to various types of outlandish and event violent cures (ibid.). For example, Maysoon Zahid, who has cerebral palsy, has talked about being dunked in the red sea multiple times as a child in the vain hope of a cure (Wong 2020).

The third belief, however, draws not on religion, but on "science," and assumes a kind of medical determinism, which assigns medical explanations to the causes of disabilities (Rohwerder 2018, 5). It is this third belief that can be linked to killings in countries in the Global North. For example, Sheila Jennings (2013) notes that "disabled girls are assumed to be different from, and inferior to those who are seen to comply with the criteria of so-called normative girlhood, they are perceived as lacking autonomy and agency" (ibid.). This belief can lead to views of a "deadly vulnerability," in which caregivers feel motivated to "end the suffering" of their disabled loved ones (see Jennings 2013; Purcell 2014). Killings of or medical treatment without consent from disabled daughters in the United States, Canada, England and Australia, such a Tracy Latimer, Ashley X, Eve, Jeanette and Marion, are justifiable as means of protecting vulnerable disabled bodies (ibid.). In all three of these biased beliefs, the association of disability with an unfortunate life is strong.

In contrast to these biased beliefs, Brigitte Rohwerder (2018) has pointed out that there are many communities in African countries which view having a disability with a positive light. For example, some communities in Chagga in East Africa view those with physical impairments as being able to pacify evil spirits. Similarly, those in Benin are often selected as law enforcement personnel. Third, children with disabilities in Turkana of Kenya are perceived as gifts from God and deserving of excellent care. And finally, many in Uganda and Kenya in more inclusive communities have thought that people with disabilities should be able to take their place as leaders within the community.

What was particularly interesting about the treatment of people with disabilities in developing countries, though, was the source of these negative beliefs and attitudes. Rohwerder summarizes: the source was from "colonialism and the introduction of medical or charity models of disability by outside actors" (ibid., 12). The biased beliefs, then, came from the colonizing "West."

The connection among disability, misfortune and charity that is present in both McMahan's examples, the fears of the transhumanists, and the just reviewed three biased beliefs has a colonial heritage. This link finds an origin in the charity model of disability, which aims to elicit pity for those who are deemed "unfortunate." The charity model invokes normative time and colonialized space in order to evoke a strong affective response, usually from donors in a capitalist economy.

Groups such as Human Rights Watch often appeal to the charity model and show images of young women or girls to vie for attention in competition with thousands of images in various media outlets (Goggin 2009; Kim 2011; see for instance, Human Rights Watch 2012). To trigger a strong affective response, groups which espouse the charity model utilize "hyper-visible" images "when inspiring an emancipatory response to the material consequences of actually living with a disability" (Erevelles and Nguyen 2016, 3). Young women and girls captured in these images are associated with vulnerability and often invoke a combination of "peril and promise" (ibid., 4).

These images are contrasted with the neoliberal media image of the female athlete or corporate executive who is resilient, independent, and assertive as a direct contrast to these war-torn images (ibid., 4; Gonick 2006; Kim 2014; see also Hahn 1997). Examples of this strategy are present again in the discourses on Girl Power and Girls-in-Crisis: on the one hand, groups such as Girl Power idealize the self-determining individual, while, on the other hand, groups such as Girls-in-Crisis, create an anxiety about those who were unsuccessful in producing themselves this way.

What is evident in the justification of these two images is the association of misfortune, disability, and colonialized space. The charity model also filters space as a strategy by linking disability to geographic location in areas exploited by colonialism. Many times, these hyper-visible images are in "racialized territories" (Erevelles and Nguyen 2016, 8).

When thinking about disability as a kind of misfortune in space, then, two additional intersectional concerns connect in our mind: race and class. According to Jonah I. Garde, the argument for this linking together goes something like this:

> people who are affected by poverty do not have access to food, sufficient health care, education, work, and housing. This in turn increases the chance of diseases, injuries, and impairments. And again, it is argued that impairments increase exclusion and marginalisation ultimately leading to increased poverty. Poverty is thus depicted as the cause and the consequence of disability and vice

> versa. Therefore, any development policy that aims to reduce poverty has to address disability. (2016, 164)

Both positive and negative images of disability have been racialized and classed by oppressive stereotypes and biased beliefs about living with a disability. Most often, these negative images are linked to images of impoverished, racialized, and Indigenous communities in the Global South. Subconsciously, it is assumed that disability, poverty, and race are linked. In essence, they are linked to misfortune.[3]

From this perspective, what then can be done to prevent the face of misfortune or, at least, overcome it? One answer is to use normative time as a strategy: promise "hope for a better future," much like the transhumanists do. As Alison Kafer notes, groups which espouse the charity model, often turn overcoming a disability into the milestone while relegating vulnerable disability to backwardness (Kola´rˇova and Wiedlack 2016, 136; Kafer 2013). Are not genetic technologies using the same tactic to alleviate our fear of the face of misfortune? Are they not promising control over a certain future?

Even if we were to use these genetic enhancements to promote societal well-being, when we begin to fill in the details, the pathway forward to our imagined future is unclear (cf. Specker and Schermer 2017). The reason for this turns on how our beliefs shape our projected future. We envision that a longer lifespan, better memory, or greater strength would entail an increase in our overall life satisfaction. Yet, what we envision are concepts and images created by industry. Similarly, when we imagine misfortune, the images which come to mind were given to us by colonialism. Instead, I propose that the virtue of empowered self-direction is needed to help us return to our actual needs and work through the lived experience of misfortune. To do so, though, I will need to return to my own reflections and then the story of Denise Sherer Jacobson.

THE VIRTUE OF EMPOWERED SELF-DIRECTION

Because misfortune is a face of vulnerability, infertility as a form of misfortune challenged me to think differently about who I was and who I would become. Life wasn't saying to me that I couldn't be a mother. Instead, it was saying to me that there is more than one way to be a mother. Coming to this realization was not something I came to on my own. As our identities are intertwined with others, and constructed by our social and historical conditions, so is the task of reconceiving a part of our identity that has been torn apart from us. It is the task of reconceiving the *self*.

In choosing to share my story with friends and family, at different times, I found myself caught in the well-meaning gaze of pity. I felt objectified, stripped of my own personal power, and ashamed for my apparent inadequacies. The culture of motherhood viewed me as broken, unfortunate, and an object of sorrow.

Like Lanier, the culture of motherhood was saying to me "you are responsible" for your circumstances and for the circumstances of your future child. Caught in the narrative of others, I was stripped of agency and ambition. It is because the gaze places us in a lived dilemma: overcome your misfortune or risk our aversion, disgust, and contempt for your situation. Become a hero in our eyes to redeem yourself.

This dilemma is false. Not only because disability rights activists have been campaigning against being treated with pity for decades, but also because choosing to overcome misfortune by being a hero will not give you back your subjectivity. The reason it won't, I learned, is because you are still living your life according to another's person's narrative. You are living according to their conception of your *self,* which prevents leading a self-directed life (cf. Haliburton 2014, 21).

My ability to reclaim my subjectivity, exercising the virtue of empowered self-direction, began by reconceiving of *my* self in *my* personal life story. This endeavor began when my therapist told me to be open to the unknown. And to do so, required that I develop a specific sense of wonder.

Two Features of Empowered Self-Direction

The emotional distress, self-doubt, and silent turmoil I felt daily about my own ability to be a mother was quieted in finding empathetic connections with others and feeling a sense of shared wonder in their life stories. In "Alternative Motherhoods," by Denise Sherer Jacobson and Anne Finger, Jacobson describes her experience of the adoption of her son David as a woman with cerebral palsy (Jacobson and Finger 2007, 142). A prominent leader in disability rights and a member of Berkeley's Center for Independent Living, Jacobson shares the stigma she faced as a mother with a disability choosing to adopt a disabled baby (ibid.). Jacobson's *The Question of David: A Disabled Mother's Journey through Adoption, Family, and Life (1999),* challenges what it means to be a "normal" mother by rethinking motherhood, "maternity and disability alike" (Jacobson and Finger 2007, 142). The work also critiques "a cultural tendency to perceive both of these experiences as solely private matters, thus ignoring the politicized social relations that inform disabled women's lives and identities" (ibid.). In her story, she is forthright about the conflicting emotions and judgments about her own capabilities while simultaneously challenging the patriarchal and ableist ideas about motherhood.

The myth of the "normal" mother idealizes the "soft, fleshly 'naturalness' of breasts, arms, and laps—that most inspires resistance to disabled women as mothers." (ibid., 137). Many disabled mothers "desire to do the ordinary," which is often viewed as routine. To have such a desire, though, is deemed self-indulgent, irresponsible or capitulating to patriarchal ideals by the non-disabled (ibid.). These stereotypes and judgments lead to "birth control" in our heads, the doubts and fears that any of us will ever be "decent moms" (ibid., 143–44). Jacobson herself tells her story openly and honestly about how she wrestles with the doubt that she will "be a good mother" (ibid., 145). Through her openness, Jacobson reclaims her subjectivity that the culture of motherhood tried to take away. In doing so, she exhibits the virtue of empowered self-direction.

The virtue of empowered self-direction is cultivated when we realize that the self is a product of social relations. When I think of my *self* as my subjectivity this means that I find a sense of meaning from the way I construe my life events into my life story. Without empowerment, I risk falling into the expectations of another. Further, I risk losing my subjectivity to another's storyline. I risk becoming an object in another's narrative.

As I discussed in chapter one, the origin of empowered self-direction finds its home in Young's essay called the "City of Difference" (1990b). In this essay, she defines empowerment as the agent's participation in decision-making through exercising one's voice and vote. Extending the ideas in her essay, I argue that the first feature, which is foundational for empowered self-direction, is a sense of wonder.

What I mean by wonder begins with a respectful stance of openness. Young's model of "differentiated solidarity" includes the value of openness, which includes the recognition of another's difference (1990b, 251). Being open to others means being open to bodily diversity and listening to the other. Garland-Thomson develops this openness to diversity as follows,

> [s]eldom do we see disability presented as an integral part of one's embodiment, character, life, and way of relating to the world. Even less often do we see disability presented as part of the spectrum of human variation, the particularization of individual bodies, or the materialization of an individual body's history. (2005, 1568–69)

This includes an openness to ourselves as well as openness to the needs, interests, perceptions, or values of others (Young 1997, 358). It includes a sense of mystery at one's *self* and another: it is "being able to see one's own position, assumptions, perspective as strange, because it has been put in relation to others" (ibid.). When my therapist challenged me to be open to the unknown, she meant that whatever judgment I was receiving did not matter. Further, she was challenging me to recognize that my vision of motherhood along with my life plans and choices would remain "small" if I bought into the ideology about motherhood designed for capitalist consumption.

Second, to develop empowered self-direction means to develop particular agentic skills which help us reclaim our subjectivity. Specifically, these agentic skills are based on our interdependence with others and not an atomistic notion of autonomy. Feminist disability studies has a critical strategy to question the assumption that disability is a flaw, lack or undesirable trait one needs to eliminate. In this strategy, it teaches us the skills to de-link disability and the lived experience of misfortune.[4] Further, it teaches us how to understand our subjectivity as interconnected with social and cultural meanings.

To reclaim our subjectivity means we need to de-link the cultural biases and stigmas from the cultural meanings surrounding the human body. Garland-Thomson outlines a critical strategy feminist disability studies uses:

> It questions the assumptions and negative attitudes linked to biased terms used to describe disability, such as "deformity" or "abnormality," and instead seeks to use precise language such as "the traits we think of as disability." This strategy can help to break down the stigma surrounding disability and to call into question what is meant by terms such as "happiness," "attractiveness," "suffering," "dignity," or "a livable existence." (2005, 1568–69)

Moreover, it calls on us not only to change our societal practices, but also to question our beliefs about one's life plan and sense of self. What this meant for me, for Lanier, and for Jacobson, was to jettison the capitalist culture of motherhood and pregnancy. It meant not letting others dictate our life stories and strip us of our subjectivity. And further, it meant empowering us to reconceive of what our family lives could be.

Application to Facing Misfortune

The face of misfortune taught me personally the virtue of empowered self-direction. While I had always been independent, the idea of autonomy could not help me deal with the lived experience of this face of vulnerability. The reason for this limitation turns on the fragility of our future. It involves coming to grips with being caught in "bad luck."

Autonomy, to me, has always involved agency in an atomistic fashion. To be autonomous meant I was somehow responsible for my actions and choices. And yet, to look for someone or something to be at fault in the lived experience of misfortune is to confuse outcomes for agency. I think this is why misfortune is not a popular topic in ethics: how can we assign blame when there is no one at fault?

This is why I found that recovery could come neither from hope nor acceptance, because those feelings cannot extricate us from the ideological narratives others assign to us. Rather, what I learned from my therapist, the adoption agency, and others was that we reclaim our subjectivity by letting go. We are able to do this when we recognize that our vision of what makes for a good life was too small.

In her memoir, Lanier recalls the sermon her husband gave for Lent. Lent is about giving something up and it should hurt. It is about surrender. Justin talks about how maybe you try to give up chocolate, and that hurts. Maybe the next time you give up the belief that you're right in an argument. This hurts a little more. And then you give up a grudge against a family member—and that really hurts (2020, 297). For Lanier, it was giving up the image of a SuperBaby:

> Seven years earlier I had to give up the kind of child I thought I was supposed to have. Not just the SuperChild. Not just the child who hiked, scythe in hand, the preferred upward slope of the bell curve. I had to give up the child I could assume would outlive me. I had to give up the child I knew for certain would grow beyond her need for me. I had to give up the child who would 100 percent walk. That would 100 percent talk. Some of those surrenders would come back to me (2020, 297).

As she notes, maybe we give up "so we can more fully receive" or so we can receive "something better than our small minds had wanted" (ibid.). According to Lanier, Fiona did not teach to not take the little things for granted but rather teaches that "the things I've considered quotidian are also miraculous" (2020, 299).

Fiona taught her parents the virtue of empowered self-direction. As Lanier and Jacobson have shown, having a child with a disability has nothing to do with an unfortunate life. Children with disabilities are often the first to challenge their parents to develop the virtue of empowered self-direction.

For example, for children with mental health differences, "parents experience a deepening sense of self-awareness and inner strength" (Aschbrenner et al. 2010, 606). According to Kelly A. Aschbrenner et al., caregivers have reported "becoming stronger, more tolerant, less judgmental," as well as gaining sensitivity and empathy toward others (ibid.). A fact of life is that we all are born into a series of familial, social, cultural, and now even international webs of connections (Baier 1987, 53). Empowered self-direction as a virtue recognizes the web of interconnected, and sometimes conflicting, relationships that shape our identities and individuality.

For empowered self-direction in practice, I think that the philosophy behind the theory of DisCrit proposed by Subini Ancy Annamma, David J. Connor, and Beth A. Ferri provides way to understand this virtue in an educational setting. According to Annamma, Connor, and Ferri, a "DisCrit theory in education is a framework that theorizes about the ways in which race, racism, dis/ability and ableism are built into the interactions, procedures, discourses, and institutions of education, which affect students of color with dis/abilities qualitatively differently than White students with dis/abilities" (Connor, Annamma, and Ferri 2016, 14). There is a sordid history of educational segregation based on race, class, ability, and nationality. The philosophy of DisCrit seeks to overturn these practices with inclusive pedagogy and privileging "voices of marginalized populations, traditionally not acknowledged within research" (ibid., 19–29).

The aim for this theory, then, is not just some ideal form of enhanced human happiness, but an inclusive ideal, which considers the completeness and diversity of human life (Becker 2012). Clare captures the openness and agency within empowered self-direction found in the philosophy of DisCrit:

> These moments allow us to turn away from *normal.* They interrupt the many ways shame hooks us into a cure. They create space for our body-minds as they are right now. They foster a matter-of-fact acceptance of our tics, tremors, stutters, seizures, knots, scars, pain, quirks. They encourage an appreciation for missing teeth and the smarts that stretch food stamps to the end of the month, for big bellies and wide hips, for the flash of hands signing American Sign Language and typing on assistive communication devices, for dark skin and kinky hair. They let us embrace our wild femmeness, our handsome butchness, our glorious androgeny. I want to live in a world where these moments are common and unremarkable. (2017, 167)

To shape the possible lives of our future children, we do not need to shape their bodies (Campbell 2015; Parens 2006). Instead, we need to shape our present societal practices.

The virtue of empowered self-direction can play out in time and space with the right inclusive pedagogy. At the end of her memoir, Lanier talks about Fiona's school, which uses inclusive pedagogy like DiscCrit. At school, Fiona's teacher told Lanier about an activity in which students introduce themselves sharing their favorite color using a talking feature on an iPad. This device looked similar to Fiona's talker, so Fiona felt like she fit right in. Lanier describes the rush of emotion she felt hearing about the activity that enhanced both her daughter's learning and her peers' learning because the app represented both the picture and the word (2020, 269–70). In this shared activity, the students practiced openness to different styles of communication.

In Fiona's story, we can witness various "agentic skills" such as self-discovery, self-definition, imagination, and introspection in order to develop empowered self-direction. These skills can be either developed in supportive environments or hindered by oppressive socialization (Meyers 2005). The skills include the development of our emotions, desires, imagination, and personal values of concern, which should be viewed within the context of one's social environments.[5]

We can witness this shaping of our social practices when Fiona's speech therapist introduced three ways the students could say "good morning" to each other. They could use their words, their ASL signs, or they could use technology by using the app on the iPad. This communication system was inclusive and communicated that other forms of communication are valid. It also meant, as Lanier noted, that Fiona was a valued member of the community (2020, 288–89).

Coming back to where we began, being open to human variation, diverse perspectives, and differently imagined futures would be a better foothold from which to evaluate these genetic technologies, which promise to give us a certain future within our control.[6] A future that has been assumed we could choose. I have to question, however, whether the use of genetic technologies for the improvement of various skills and abilities provides a better future and an enhanced life. In our present society, we cannot promote the kind of utopia Bostrom and other transhumanists hope for without "making schools, public places and work environments generally available to very large classes of people with disabilities" (Becker 2000, 62). We have to change our global society first, because if we do not, we risk making the wrong decisions about our possible future. Further, we risk commodifying human embodiment and well-being. And most of all, we risk limiting our conception of the good life, rather than expanding it.

CONCLUDING THOUGHTS

When my husband and I announced we were adopting, we received mixed responses. Some friends and family were filled with joy; others gazed upon us with pity. And finally, others praised us for doing "a good thing." What was meant was that, in their minds, adoption was an act of charity for an "unfortunate" child.

I feel fortunate that I didn't buy into their narratives about our life. I feel lucky to have been blessed with the wisdom to know that a child is a good, and that it is foolish to think otherwise. Those who hold such as view are subjecting their conception of the good life to the focusing illusion. Whether with visions of a technological future or our evaluation of the daily minutiae, our imaginations tend to lead us astray.

As a virtue, empowered self-direction taught me that wonder is what provides the basis for our possible life plans and choices. Further, it aids in the understanding of our consequences for acting from our experiences and interests. Our empowered self-direction is interlocked in the narrative histories and interests of others. The life plans we choose to purse are meaningful, not because of some image given to us by capitalism, but because they are interconnected with the choices and plans of others. Choosing a life path is not an atomistic endeavor. Instead, it is viewing one's life as interdependent, social, and inclusive.

In the final chapter, I address suffering as a face of vulnerability and return to Kyle's story. In this chapter, I address the challenges genetic technologies have for eliminating suffering and why having a disability does not entail having a life not worth living. In response, I propose the virtue of mutual recognition as a guide for this face of vulnerability.

NOTES

1. Stan van Hooft argues that Nietzsche considered pity to "belittle the person who feels it as well as the person who is its object. It belittles the person who feels it because it shows that the person fails to appreciate the positive power of suffering, and it belittles the object of pity because it represents that person as failing to bear suffering courageously" (1998, 18).

2. According to Buchanan, biomedical enhancement can include any of the following modes: "engineering of human embryos and gametes (the insertion naturally occurring genes, either from humans or from insertion of artificial chromosomes or products of synthetic and the manipulation of gene regulatory functions); enhancements (administration of drugs that affect body); human-machine interfaces; and laboratory-tissues (presumably using stem cell technologies and synthetic biology). Kinds of enhancements include improved affective, and motivational capacities; increased longevity; resistance and/or complete immunity to various diseases. Motivational enhancement drugs already exist, and aging predict that significant extensions of life will possible, through regeneration of tissues and organs employing technology or, more radically, by slowing or even arresting of cell senescence" (2009, 350).

3. Similarly, there has been a medicalization of race, class, and disability. In *Dark Ghettos: Injustice, Dissent, and Reform,* Tommie Shelby argues that a medical model has been applied to race and class as a prescription to "increase the material welfare of people living in ghettos" (2016, 2). As Shelby says, "policymakers working within the medical model treat the background structure of society as a given and focus only on alleviating the burdens of the disadvantaged," but it is "the technocratic reasoning of the medical model" that "marginalizes the political agency of those it aims to

help" (ibid.). Misfortune, here, is linked by both medicalization and charity through the medical gaze, which uses the social frame of normalcy to determine which bodies are healthy.

4. Some of these narratives people with disabilities are placed in, Rosemarie Garland-Thomson notes, include the following characteristics: (a) overcoming a personal defect rather than viewing disability as bodily transformation, (b) a catastrophe "that presents disability as a dramatic, exceptional extremity that either incites courage or defeats a person" and (c) disability presented as something that "one can and must avoid at all costs" (2005, 1568).

5. Skills such as these are encouraged at schools like the Birchwood program in New York at the following link: https://www.ccsd.edu/domain/487

6. Empowered self-direction must not be curtailed by bias and fear regarding genetic technologies. Medical counsel is often guilty of excluding the perspectives of people with disabilities. Cognitive disability, for example, has been excluded from discussions of autonomy, while the lived reality of the disability, the nature of the disability and the historical basis have not been taken into account (see Kittay and Carlson 2001). According to Arneil, "the image of 'autonomy' of the nondisabled is built upon a conceit at best, a false image at worst, because it is only possible when various kinds of interdependent social arrangements or 'accommodations' are already present" (2009, 236). There is often a conflation of physical impairments with mental impairments; both are excluded from traditional accounts of autonomy (ibid., 221). Instead, the discussions between counselors and parents should produce wonder and should conceive of the future child's possible life plans from an inclusive and informed perspective. To be inclusive and informed, advice given to future parents should include a diverse representation of perspectives on living with visible as well as invisible disabilities (Davis 2005).

Chapter 5

Facing Suffering

Capitalism, QALY, and Well-being

INTRODUCTION

"You'll always have a home with us," Kyle said as he stretched his thin legs out on my future mother-in-law's teal couch. When I was growing up, I changed schools a lot. While adjusting to new school settings every few years taught me how to make friends easily, it also taught me the pain of losing them on account of distance. Changing schools at times made me feel adrift in the world, like a solitary traveler, who never knew where to hang her hat. I never quite knew who "my peeps" were. Kyle was the person who changed that for me.

Kyle and my husband had been close friends since elementary school. With his blond hair, blue eyes and charming smile, Kyle had inherited his father's good looks. But he had not inherited his father's body. Kyle was born into a family of athletes. As a child, he grew up playing many sports, and like his father, he was quite talented at them. By age twelve, due to Kyle's progressing health condition, he could no longer play sports with his family and friends. By age twelve, Kyle like me, was set adrift in his social world.

While his condition set him apart, it also enabled Kyle to develop a kind of maturity about relationships at an early age. He, too, was now a traveler like I was. His bodily change forced him to drop out of sports and to find new friends. But as Kyle says, this bodily change helped him become a better person.

His condition enabled him to choose his own friends—friends who saw Kyle the person, not Kyle, the boy with the "broken body." One of those friends that Kyle chose was my husband. In choosing his friends by an act of fate, Kyle was able to dictate with whom to share joys, cries and laughs through junior high and high school. Unlike other teenagers, Kyle did not feel any pressure to fit in. Because he never would. And because he never felt that pressure, he was well liked in high school.

I asked in the introduction of this book if Kyle's parents had known about his condition, would they have chosen to terminate the pregnancy of Kyle. Would they have been advised to terminate Kyle, because to give birth to him would be a harm? Would it be because Kyle was going to have "a life not worth living?"

Some philosophers, including Singer, Buchanan, Harris and Purdy, would answer this question in the affirmative.[1] The most prominent reason, or cluster of reasons, a couple (or single reproducer) may want to employ genetic engineering for their future child is to prevent the difficulties and suffering of a particular disease or disability. Most agree that avoiding a severe or painful disease, such as Tay Sachs, or a severe disability that would result in *a life not worth living* is not controversial (Glover 2007, 52). But what is meant by the phrase "a life not worth living"?

I think what these philosophers and potential parents have in mind is that they do not want a future child to experience the final face of vulnerability if they can prevent it: the face of suffering. In our minds, when we imagine suffering, we think of a life in constant physical pain and mental anguish, a life filled with insurmountable obstacles, and reaching our psychological limit of being unable to cope. The life we envision is a life devoid of peace and solace.

And yet, the lived experience of suffering runs on a continuum. This continuum is hard to pin down. We all are vulnerable to suffering physically, psychologically, and morally at some point in our lives (May 2017, 4). And so, when we try to quantify an aspect of lived experience that is personal, and to some degree particular to each person, we arrive at the phrase "quality of life." The reason that these philosophers say the pregnancy of Kyle should have been terminated is because they view people with disabilities, especially those like Kyle, as being "worse off," that is having little to no quality of life (Lanier 2020, 275–96). And further, these philosophers argue that genetic technologies such as enhancements should be pursued because they prevent lives not worth living from happening in the future.

I disagree with their arguments for both personal and philosophical reasons. It does not necessarily follow that the presence or absence of a trait predicts the quality of life a child will have (Johnson 2020, 9). To experience pain one day, does not mean that one cannot experience joy the next. This is because suffering as a face of vulnerability targets our lived experience and particular circumstances—it does not target our biology. Rather, the lived experience of suffering is a facet of the fragility of the human condition.

In this chapter, I address the bias of those philosophers who think that genetic technologies can grant us a future with much better lives guided by procreative beneficence. I then argue that reliance on measures such as QALY not only marginalizes voices and misrepresents experiences, but also commits an epistemic injustice. The virtue needed, instead, is that of mutual recognition, because people with disabilities have a privileged perspective on well-being that should be valued. To begin, though, it is necessary to consider the face of suffering, because it is the face philosophers consider when they deliberate on whether the disabled should live or die (Lanier 2020, 260).

SUFFERING AS A FACE OF VULNERABILITY

The face of suffering need not show itself to us as physical pain, and yet often, that is how we imagine it. Sometimes, when we think of suffering, images of writhing in pain, being unable to move without the assistance of others, being fed by another, held by another, soothed by another, and having to be changed by another come to mind. The shame of our bodily dependency and physical agony are too much to bear. And we fear being a burden on the ones we love.

There is another side to suffering, though: that of mental anguish. Brought on by biology or circumstance, suffering shows itself as grief, guilt, sorrow, anxiety, or trauma. At night, the affliction of the mind will not let us rest. If we do sleep, we are only to awake in terror. The social shame and rejection from others cause us to slip out of view, whether into bed or onto the bathroom floor. The shame of the mistakes we've made have been too great. The losses we've had have been too big. And we fear we're only holding our loved ones back.

The lived experience of suffering grips the existential dimension of our lives. Suffering comes in the night and haunts us. It is the guilt that won't let go. It is the pain that will not numb. It is the agonizing distance felt from those to whom we were once close. Suffering shows its face as isolation, as guilt, as rage, as shame—as the mocking which won't end. The child in us ceases to cry because we know that comfort will not come. And yet, we are still waiting.

We fear the face of suffering because we envision a life not worth living. In our minds, this is a life without joy, without love, and without laughter. A life in constant physical pain and mental agony. It is a life devoid of hope and yet, the emotion we feel is not despair. No, it is a life which feels empty, meaningless, and alone. We fear the crisis of meaning that is also constitutive of the experience of suffering (cf. Wendell 1996, 173).

In normative time, the lived experience of suffering is attuned to the present. As a state of being, it disrupts our daily tasks, casts doubt on our hopes for happiness, and alienates the spiritual dimension of our existence (van

Hooft 1998, 13). In normative time, we think of our lives as one single narrative unity from birth to death. In our imagination, we envision ourselves as having a perceived future, a distinct personality, past experiences and memories, cultural background, behavior, and relations with others. Suffering disturbs all dimensions of our lives. It is pervasive.

In normative space, the lived experience of suffering suffocates us or evicts us from where we feel at home. Our place becomes strange, and we cannot rest or find our way (Bueno-Gómez 2017). As a lived experience, suffering can include our loss of self and an "unhomelike being in the world" (ibid.). Suffering alienates us from our own body. It alienates us from our engagements in the world with others; it alienates us from our life values (ibid.). It makes us strangers to ourselves because it can impede us in living the lives we wanted. It is unhomelike insofar as we exist in an uncomfortable way in an uneasy environment without rest (ibid.).

We think of suffering as the constant aching in our muscles: in our arms, our chest, our back (Wendell 1996, 172). We try to concentrate on other things or drift away to that safe hideaway in our minds where no one else can go. When we envision suffering, we envision pain. And to escape from the pain, we retreat from the present. To escape from pain, we retreat from *now* and we retreat from *here.*

Suffering becomes part of a person's life. It is a part of our lives that we cannot control (May 2017, 88). Genetic technologies, however, promise the prevention of suffering. They offer a way to protect ourselves and our future children from a life not worth living. They give us the hope that everyone in the future can have a quality of life. With science as our solution, then, philosophers have argued that enhancements give us the opportunity to eliminate our biological precarity. And the first target of eliminating a life not worth living are traits for disabilities.

GENETIC TECHNOLOGIES TO PREVENT A LIFE NOT WORTH LIVING

"He insists he doesn't want to kill me. He simply thinks it would have been better, all things considered, to have given my parents the option of killing the baby I once was . . ." writes Harriet McBryde Johnson in reference to Peter Singer (Johnson 2020, 3). Johnson, who had spinal muscular atrophy, notes that Singer wants to prevent the suffering that "comes with lives like mine" because parents prefer healthy babies (ibid., 10).

Some have argued that the goal of medicine should be to alleviate suffering (Haliburton 2014, 230). Alleviation can come in the forms of reducing physical pain, aiding psychological comfort, and even using assisted technologies.

Philosophers like Singer, however, want to go further. In general, they want to prevent suffering. And if they cannot prevent it, then they want to end it as soon as possible. To prevent suffering in the future, some philosophers have turned to the principle of procreative beneficence to guide genetic technologies. Their goal is to prevent the birth of children like Harriet and Kyle.

Procreative Beneficence and Preventing Disabilities

Some philosophers have argued that genetic enhancement could benefit society and give everyone a life worth living (Bostrom 2005b; Bostrom and Sandberg 2008; see also Tremain 2017). These enhancements would increase human capacities in order to increase total aggregate happiness and well-being. For example, equipping humans to have stronger immune systems will not only increase individual quality of life, but it will also decrease the burden on society for treating so many devastating illnesses such as Werdnig-Hoffman disease, a type of spinal muscular atrophy.

The overall benefits outweigh the potential risks and uncertainties. If we have the ability to reduce the costs of medical resources and prevent medical futility by reducing the number of expensive and painful health conditions, then are we not obligated to do so? Would this not be the ideal solution for both parties?

Transhumanists, such as Bostrom, Julian Savulescu Anders Sandberg and Guy Kahane seem to think so. They argue that we should aim to give the future child "the best life possible," which would include using genetic enhancement technologies. According to Savulescu, we should act under the principle of "procreative beneficence," which holds "that couples (or single reproducers) should select the child, of the possible children they could have, who is expected to have the best life, or at least as good a life as the others, based on the relevant, available information" (2001, 413). Thus, if a health condition would lead to disadvantages in life and a decrease in well-being, then we are obligated to utilize these technologies to prevent it. Kyle's health condition would probably qualify in the eyes of transhumanists. In short, for transhumanists, Kyle like Harriet, should never have been born.

In theory, transhumanists argue that the potential for true improvements in human well-being and flourishing are only attainable through technological transformation, which can yield tremendous benefits. Savulescu, however, does agree that deaf parents should have a deaf child if they so desire. He states:

> If they strongly value having a deaf child, and have great loyalty to the Deaf culture, it is a "ground project" to have a deaf child, then they may have most reason to have a deaf child just as [a] husband might have most reason to save

> his wife rather than two children on the basis of his present pattern of concern. (2009a, 234)

What is important for the deaf couple is that there are no serious resource constraints and that no one is harmed in the process. Savulescu claims that "other things being equal, we should save and create beings with longer and better lives" (ibid., 241). This is similar to organ transplantation insofar as we give organs to those most in need for the purpose of giving them a better chance at a longer, better life. But parents should still retain their liberty to choose. Even if the value of their future children would be considered inferior, the parents should retain the liberty to procreate their line (ibid.). Savulescu continues: "[h]owever, it is plausible to claim that they should have silicon children, rather than carbon-based children, if these children are better and are expected to have better lives just as couples should have hearing rather than deaf children" (ibid.).

What is meant by "the best life" in theory is merely the life realized by the addition of human enhancement technologies. We refer to and know about how well-being is increased from various social scientific studies supporting various theories of well-being whether hedonic, desire-satisfaction or objective lists. These studies show that certain elements make our lives go better and any one of them could work.

While Savulescu has argued that he has not committed himself "to any particular substantive conception of the good life," he does include particular capacities and elements which provide a mix of hedonic and objective list accounts of well-being (2007, 286). For example, in his essay "Autonomy, Well-Being, Disease, and Disability," Savulescu states that he considers well-being to be constituted by hedonic states such as pleasure and the absence of pain while also engaging in objectively valuable activities such as developing one's talents, gaining knowledge or having rewarding personal relationships (2009a, 64).

In practice, however, these philosophers seem to have in mind qualities of comfort and security every time they list a capacity to be enhanced or try to define disability. They seem to want to eliminate any traits that could cause "discomfort." For the transhumanists, as a result, I think that they would argue that genetic technologies should be used to prevent the birth of people like Kyle and Harriet from being born.

And I think they would do so in practice based on assumptions they hold for a good life and capacities of comfort. For example, Savulescu provides a welfarist position, and argues that most people care about "the human properties of value: empathy, love, wisdom, etc." (Savulescu 2009b, 234). Additionally, he has argued that we have a "moral obligation to have the best child, that is the child with the best opportunity of the best life. Morality requires that people have healthy rather than disabled children" (ibid., 240; Savulescu 2001). For example, if parents using in vitro fertilization (IVF) have a choice to select one of two embryos which are genetically identical except that one embryo carries a gene with a predisposition for asthma, then according to Savulescu's principle of procreative beneficence, the parents should select the embryo without the risk of asthma (Bostrom 2005a, 19).

Savulescu, though, seems to be aware of the vagueness of his definitions for "human enhancement" and the bias for his conception of "disability." In his article, "Justice, Fairness, and Enhancement," he admits that the term human enhancement is "itself ambiguous" and that when transhumanists consider it, they mean "some change in a state of the person—biological or psychological—which is good" (2006, 324). Changes that are good are what should be promoted or maximized. For Savulescu, the "value in question is the goodness of a person's life, that is, his/her well-being" (ibid.). Savulescu then defines his welfarist definition of human enhancement: "Any change in the biology or psychology of a person which increases the chances of leading a good life in circumstances C" (ibid., 325). To enhance is to increase the value of a person's life whether through medical treatment of disease, increasing natural human potential or superhuman enhancements (ibid., 325–26).

Further, Savulescu has proposed an interesting account of the relation of disability to well-being in his essay "Autonomy, Well-Being, Disease, and Disability." In his essay, he defines a normative account of disability, which includes some elements of social construction:

> a relatively stable physical or psychological condition X of person P counts as a disability in circumstances C if and only if X tends to reduce the amount of well-being that this person will enjoy in C. This is a welfarist account of disability that relates disability to well-being. Disability, a normative term, is not the same as inability. The inability to wiggle one's ears is not a disability. (2009, 63)

What is of interest is the case he cites, that of Ashley the so-called "Pillow Angel," which he uses to illustrate how his definition differs from the "normal function" or "species-typical functioning" account of embodiment. For Savulescu, according to his welfarist definition, the treatments Ashley received would be considered enhancements because they improved her overall well-being (ibid., 63–64).

There are two points of interest, here, that I would like to consider. First, in reference to Ashley, on her parents' website is the following description: her "cognitive and mental development level" will "never exceed that of a 6-month old child," and it is unlikely that she will ever "be able to walk or talk or in some cases even hold up [her] head."[2] Alison Kafer, in *Feminist, Queer, Crip*, has argued that the rhetoric behind the term "pillow angel" both "reflects and perpetuates this linking of disability with infancy and childhood" (2013, 52). Ashley's parents call her their "Pillow Angel," according to Kafer, because "she is so sweet and stays right where we place her usually on a pillow" (ibid.). Kafer notes that "this phrasing paints a picture of infant-like dependency and passivity"; it makes it difficult to imagine Ashley as "a teenager or a woman-to-be" (ibid., 55). This linking of disability with infancy and childhood has a problematic history of linking adulthood to independence, autonomy and productivity, something that would be unachievable for Ashley (Purcell 2016b).

Second, Savulescu indexes his conception of disability to well-being as an impediment. While it is possible that having a disability can disrupt one's life and make it more difficult at times, Elizabeth Barnes has argued that having a disability does not overall make one's life worse and does not make it more probable that it will (2009a, 339). Instead, we should note that the inference from disability as a harm to disability as a negative difference-maker stands in need of logical support (ibid.).

Rather than define disability as an impediment to our well-being, it would be better to consider the actual societal impediments to well-being which structure our choices. We have to remember that our "choices are structured by oppression" (Johnson 2020, 20). Instead of trying to prevent people with disabilities from being born, as Johnson argues, we should combat the actual causes which interfere with having a good quality of life, such as "dependence, institutional confinement, [and] being a burden" because all of these are "entirely curable" (ibid.). Further, these causes can be the target for social justice in particular circumstances.

Philosophers in response, though, turn to a phrase to make their theories work: *a life not worth living.* This life is a theoretical life, and I argue, is an empty concept. When forced to provide a concrete example of what they mean, philosophers in response will rely upon the phrase "quality of life" (QALY) to justify the preference for some lives over others.

QALY and A Life Not Worth Living

Any time there is a disability related article on the Internet, we only need to scroll through the comments to find someone saying people with disabilities should die (Lanier 2020, 260). Similarly, we can witness newspaper reporters extend sympathy to parents who killed their children who had disabilities. Often, the parents of these children have been told that their children would have little to no quality of life. These children, Heather Lanier notes, are like her daughter Fiona (ibid., 296). These children, too, are like my friend, Kyle.

In *Foucault and Feminist Philosophy of Disability*, Shelley Tremain argues that as "predictive testing strategies are directed toward progressively earlier and earlier stages of a pregnancy, along with the fact that in vitro fertilization (IVF) and preimplantation diagnosis (PGD) have become more and more widely available," there is reason to be gravely concerned about the motivations to develop this technology in order to eliminate what is viewed as a defect (2017, 167). The messaging behind these technologies is that "disabled people's lives are not worth living, nor worthy of support" (Tremain 2017, 167; cf. Rock 2000). The messaging is that their lives should be ended as soon as possible.

To answer to the question I posed in the introduction, some philosophers such as Singer would have advised that Kyle's parents terminate the pregnancy. Singer's view of living with a disability is not unique or isolated. As Johnson has described, people have said to her that if they had to "live like you, I think I'd kill myself" (2020, 7). She responds by saying:

> I used to try to explain that in fact I enjoy my life, that it's a great sensual pleasure to zoom by on my power chair on these delicious muggy streets, that I have no more reason to kill myself than most people. But it gets tedious. God didn't put me on this street to provide disability awareness training to the likes of them. In fact, no god put anyone anywhere for any reason, if you want to know. But they don't want to know. They think they know everything there is to know, just by looking at me. That's how stereotypes work. They don't know that they're confused, that they're really expressing the discombobulation that comes in my wake. (ibid.)

It is quite common for those with able-bodies to question the happiness of those with disabilities. This questioning can even cause us to doubt our perspective of happiness. Yet, what I have learned through my own explorations is that these prejudices do not constitute an "objective perspective" on quality of life.

Some philosophers, though, do argue that they do have an "objective perspective," and Singer is among these. They refer to the quality of life metric as a reason to discriminate against people with disabilities. Using QALY, evaluators are asked to consider the overall "burden" of living with a disease or decrement in health functioning in order to rank the "quality of life" (QALY) (see also HRQL "Health-related quality of life). The cost-effective allocation of medical resources, and future medical resources such as enhancement technologies, is commonly used for the aggregate benefits of those within society.

When evaluators use such measures, they have three guidelines in mind. These main guidelines are often assumed when making a decision: "(1) a broad 'rule of rescue' that gives priority to those in immediate need, (2) health maximisation and (3) equalisation of lifetime health" (Cookson and

Dolan 1999). These principles generally rest on various different priorities such as the maximization of general population health, the reduction of health inequalities, and to respond to life-threatening situations. Fundamentally, though, these principles must always be guided by practical and budgetary constraints (Baltussen and Niessen 2006).

A common objection to the above view is that it can lead to inequitable treatment for people with disabilities (Wasserman et al. 2005; Bognar 2011; Cureton and Wasserman 2020). There are two issues raised against using QALY or HRQL. First, this measure falsely assumes that disabilities negatively impact an individual's well-being due to the negative impact disability has on social participation and daily living. For example, as John Harris notes, QALY has tendencies to be ageist because QALY values life years rather than people's lives and to place people with disabilities in a kind of "double jeopardy" (1987, 119–20). Similarly, disability scholars are quick to point out that QALY's lower ranking of the benefits people with disabilities might receive from medical resources exaggerates the negative impact of living with a disability.

Second, QALY overlooks the social and environmental factors that limit those valuable activities. For example, when we consider life expectancy, in stable nations with good medical systems "many people with major physical and cognitive disabilities have life expectancies after onset that are 85 percent to 95 percent of that of the population as a whole" (Becker 2005, 10). Lawrence Becker argues that this rise in life expectancy is because substantial social resources and necessary caregivers are available (ibid.).

Third, those using Quality of Life instruments assume that the "disease burden" placed on society will be too high rather than recognize how individual physical differences are at play (Koch 2000).[3] In other words, they forget the effect an exclusionary environment can have on individuals with disabilities. People with disabilities often suffer from stigma, discrimination and marginalization within education and the workforce (Purcell 2014).[4] If the cause is one's social environment and lack of resources, then using QALY as a metric for the distribution of resources follows from bias, and so is unfair to those living with disabilities; they should receive more resources, rather than fewer resources, to make this situation just.

The question of whether the disease burden placed on society being too high is interesting. It is of interest because it construes this debate in terms of distributive justice, that is, in terms of questions of fairness. And the way that we measure fairness in society, though, is in terms of costs and distributed material goods. It makes me wonder if there is a financial amount attached to a life.

THE EPISTEMIC INJUSTICE OF COSTING TOO MUCH

Is there a difference between a life not worth living and a life not worth paying for? After dropping out of college, Kyle left home to live in Virginia briefly before returning home to Idaho because his condition had worsened. He soon discovered that with government assistance, there were many human goods which he would be incentivized not to pursue.

When Kyle first walked into the government office to apply for assistance, he was asked "when did he decide he was disabled?" Taken aback by the question, Kyle soon learned of the financial incentives involved for allocating costs and benefits. For example, medical procedures and pharmaceuticals with significant costs through Medicaid were encouraged, while Kyle's lifestyle choices such as eating healthy food and practicing yoga were not reimbursable. Second, the office informed him that because he did not have a minimum of ten years of work experience, he did not qualify for social security benefits. Instead, he only qualified for supplemental security income, which covered basic needs for food, clothing, and shelter. Further, Kyle learned that while the maximum federal benefit is approximately $700–800 per month, this amount is reduced if one is married, lives within a family structure, earns a small amount of income, and as Kyle discovered, lives within a family member's home. As a result, Kyle's monthly check has been reduced to $529.34, for which he must reapply for every three years. Kyle's personal experience has led me to reflect on a different set of philosophical questions. Does Kyle cost too much? How much is a life not worth paying for? How much is too much?

Capitalism motivates people to desire comfort and gadgets, which come with material gain. This motivation comes from a desire for security and the avoidance of risk. We would prefer to lead lives manipulated by large corporations rather than exercise our individuality (Fromm 1961, 4). And what is a life that lacks comfort? A life that is a burden on others and society. Capitalism would call that a life not worth living. A life which carries an overall burden. Further, this would be a life we could measure. A measurement we could call "quality of life" (QALY).

As Jonson notes, Singer wants to legalize the killing of individuals with severe cognitive impairments at any age: "so severe he doesn't consider them 'persons'" (2020, 3–4; e.g., Singer 2009). It is because in Singer's mind, disability "makes a person 'worse off'" (ibid., 10). For example, then, when it comes to debates about assisted suicide, according to Johnson, the "care for assisted suicide rests on stereotypes that our lives are inherently so bad that it is entirely rational if we want to die" (ibid., 20). And his answer for why we should prevent children from being born? Because parents and adoptive parents want healthy babies (ibid., 10).

I, like Johnson, have "trouble with basing life-and-death decisions on market considerations when the market is structured by prejudice" (ibid.). In *Foucault and the Government of Disability,* Tremain notes that we can "understand how current discourses on 'good death' and 'choice' with respect to assisted suicide" tie "disabled people . . . to notions of empire and capitalism in the West" (Tremain 2015, 21). The cost of bodies has a colonized history within the global economy. For example, the bio-capital, pharmaceutical and colonializing quest for the use of chemically innocent bodies, the use of bodies for medical experiments, and the use of surrogate-mother or organ donor bodies have history of attaching prices to bodies for labor or supplies within the "global economy of debility" (Erevelles, N. 2011). [5]

What I find interesting in the literature is that when philosophers are asked to address the oppressive social circumstances for people with disabilities and unfair allocation of medical resources, they usually lean on two arguments. The first is the unfair cost burden to others in society. And the second is medical futility.

First, some argue that while medical resources may increase the quality of life for people with disabilities, the burden these additional resources place on other members of society is unfair (Daniels et al. 2000). Impairment-related resources such as assistive technology, which include wheelchairs and hearing aids, accessible city planning, which includes curb cuts, elevators and disability parking, and social services, which include pensions and rehabilitation, are costly (cf. Daniels 1994). Basically, there is a point in which some lives cost *too much* to pay for.

Aging populations are often cited as one example. Aging, rather than damage, is a multifaced "process leading to an exponential increase in mortality with time" (West et al. 2019, 879). Researchers West, DeGrey et al., note

> that regenerative therapies that repair the initial degenerative triggering events will have a considerable impact on lifespan. These theoretical predictions, combined with the power of technologies such as pluripotent stem cell-based cellular compositions, suggest that there may be a path not only to alleviate the high costs associated with the chronic degenerative diseases of aging but also to extend human life expectancy. (ibid.)

Many researchers are concerned about the aging demographic and the strain the large populations in industrialized countries will place on the healthcare systems. This will also cause financial strain. Combined, this will cause a health care crisis, which is an opportunity for innovative solutions (ibid., 880). This is the logic of capital.

Other examples cited are so-called lifestyle diseases such as obesity, cardiovascular disease, and certain types of cancer (Wikler 2002). The degree of control for diet and level of exercise has not yet been accepted into the set of criteria. Some thinkers, such as Alexander Cappelen and Ole Norheim, argue that it should be permitted "to assign a limited but significant role to individual responsibility in the rationing of health-care resources" (2006, 313). The burden placed on society for accommodating these lifestyle diseases should be a factor in allocating resources. In other words, is it not unjust to burden those with able-bodies in society with the demand to accommodate a disability which could have been prevented? The cost of supporting these lives is too high and unfair.

In some cases, the cost of a life is too much from the perspective of capitalism. And for these lives, this is where a second argument enters: the argument for medical futility. For example, "the public is used to dehumanizing disabled people and finding it acceptable to debate whether a life is worth living if you are disabled. Issues include assisted suicide, euthanasia, and withholding life support from the disabled babies or poor people in need of transplants" (Nocella II, Bentley, and Duncan 2012, 72). In cases like these, philosophers like Singer often argue that individuals should have the right to die.

While the argument for medical futility does include cost-cutting measures, what is of more interest is the Utilitarian reasoning used for making decisions for medical resources. Philosophers ask whether there are not some cases in which it is medically futile for which to distribute resources? Would it not be a better choice to contain health care costs by eliminating care for patients whose prognosis is poor and would receive little if any benefit from these resources (Orentlicher 1996)?

The use of medical futility reasoning can be found in the metric DALYs (Disability Adjusted Life Years), which is an alternative to QALY.[6] Similar to QALY, Disability Adjusted Life Years "combine information about morbidity and mortality in numbers of healthy years lost. In the DALY approach, each state of health is assigned a disability weighting on a scale from zero (perfect health) to one (death) by an expert panel" (Arnesen and Nord 1999). By using DALYs to measure disease burden, evaluators can determine the impact of the disease on one's life expectancy and one's health quality of life. They can also disregard other non-health and indirect burdens of the disease on the individual (Brock 2003). The World Bank and World Health Organization use DALYs to measure the burden of disease on a global scale.[7]

When evaluators use the QALY or DALY metric to determine "whether a patient's quality of life seems so poor that use of extensive medical intervention appears unwarranted,"[8] it is often the case that physicians rate the patient's quality of life as lower than the patient herself does (Jonsen and

Edwards 2018). Many times, bias or prejudice against those with disabilities is present. Also, the term "quality of life" is opaque.

From an empirical basis, it may appear that those who live with a disability enjoy a lower quality of life than those without disabilities. But from a subjective perspective, there is contrary evidence to the arguments based on an unfair cost burden and medical futility. Often referring to measurements of subjective well-being, those with disabilities rate themselves as happier and assess their life satisfaction considerably higher than objectively presumed. This conflict in assessment is what Gary Albrecht and Patrick Devlieger call the "disability paradox" (1999).

The Disability Paradox and Epistemic Injustice

Disability scholars contend that those who live with a disability have epistemic privilege, and it is their reports that should be taken into consideration for quality of life measurements rather than QALY and DALY. This privilege comes from the fact that empirical studies have found that patients perceive themselves as having a higher quality of life than does the general public (Menzel et al. 2002, 2149; see also Barnes 2018). For example, in response to Singer's prejudice, Johnson writes:

> Are we "worse off"? I don't think so. Not in any meaningful sense. There are too many variables. For those of us with congenital conditions, disability shapes all we are. Those disabled later in life adapt. We take constraints that no one would choose and build rich and satisfying lives within them. We enjoy pleasures other people enjoy and pleasures peculiarly our own. We have something the world needs. (2020, 11)

Living with a disability gives one first-hand experience about what it means to live with that particular disability. Stated differently, disability may not be a disadvantage in perspective: instead, it may be an epistemic advantage. When one considers a child who grows up in a bilingual home, it has been well documented that this child will most likely acquire language more slowly than his peers. After a few years, however, this "disadvantage" becomes an advantage for learning languages and building other intellectual skills. Could it be the case that those who live with a disability soon acquire other abilities and have a better insight into living with a disability than those who are nondisabled?[9]

The issue present is that disability in itself is not conceptually linked to a lower quality of life. The presumption that it is conceptually linked, which has been made by the public, is based on misinformation. Peter Ubel, George Loewenstein, and Christopher Jepson give the following example:

> If asked to imagine they have paraplegia, most people predict that it would have a devastating effect on their quality of life. Yet, there are often discrepancies between the quality of life estimates of patients and the public. For example, the general public estimates the health related quality of life (HRQoL) of dialysis at a value of 0.39 (on a scale where 0 represents death and 1 represents perfect health), whereas dialysis patients estimate their HRQoL at 0.56. (2003, 599)

The question concerns whether the lower quality of life for people with disabilities is an empirical conclusion or an *a priori* assumption. The two perspectives which create this confusion are the individual's perspective and the social policy perspective. If it is the case that a lower quality of life is an empirical conclusion (a), then lower quality of life for people with disabilities is confounded by environmental factors, which can decrease life satisfaction overall. By contrast, if it is the case that a lower quality of life is an *a priori* assumption (b), then the perspective from the public is itself biased. Further, in the case of an *a priori* assumption, the argument to maximize cost effective analysis would be inherently discriminatory. The challenge lies in trying to determine which perspective is epistemically accurate.

Those in favor of enhancements, and those who hold to a restricted use of these technologies to eliminate all disabilities, will argue that their position is epistemically accurate. But are they right? Consider, for example, the following two empirical studies:

> In a comparative study, the general public and rehabilitation workers had significantly less positive attitudes towards persons with disabilities than did a group of persons with spinal cord injury. In other studies, persons with disabilities had significantly more positive attitudes toward persons with disabilities than did nurses or members of the nursing faculty. These attitudes were expressed in terms of willingness to interact with and feel empathy for persons with disabilities. (Albrecht and Devlieger 1999, 977)

What philosophers usually contend is that individuals with disabilities psychologically adapt to their lower quality of life (Carel 2007). In this sense, an individual living with a disability is in a worse position to judge her life objectively. This adjustment may lead to an overestimation of one's life satisfaction and subjective well-being.

Further, they argue, while this subjective happiness may help the individual, it does not extend to the objective criteria for determining a higher quality of life. This difference in perception of one's quality of life may reflect the patients' adaptation to their condition.[10] Those who hold to the adaptation proposal often argue that this adaptation may reflect cognitive deficiencies, such as denial of the reality of one's illness or the "newly suppressed recognition of the nature of full health" (Menzel et al. 2002, 2149).[11]

What is evident in the adaptive preference argument is what Miranda Fricker calls testimonial injustice. As Tremain has noted, "there is no universal, timeless, and objective 'quality of life' that can be analytically separated from the contingent concrete circumstances in which people live" (2017, 173). The philosophical arguments that do not listen to the testimony of those living with disabilities and even consider positions from subjective well-being commit an epistemic injustice in speaking for another (Fricker 2007). This type of epistemic injustice to people with disabilities can be found in several forms.

First, much of the bias in the medical community turns on an unexamined sensibility of "normal" bodily functioning. As Ron Amundson points out, describing "individuals or groups as 'abnormal' is seen as marginalizing them by use of a falsely objective criterion" (2000, 33). This view of people with disabilities is the dominant outlook in analytic philosophy: it is "generally assumed that the life of a disabled person is clearly sub-optimal in the vigorous debates over whether and how it could ever be permissible to bring such a person into existence" (Barnes 2009a, 337). The genetic engineering debate now extends this stigma and discrimination to an individual's genetic makeup (genotype) (Geller 2002, 267). This bias would indicate that living with a disability leads to a lower quality of life is an *a priori* assumption rather than an empirical conclusion.

Second, as Alison Kafer notes, overcoming disability is often turned into the milestone in healthcare measurements (2013, 28). According to Kafer, the individual who can overcome one's disabled state becomes a sign of progress and proof of an effective distribution of medical resources. The bias, then, is present in the standard of measurement. People with various disabilities are compared to what is considered "normal functioning," and then the goal should be to move them as closely to the "norm" as is possible. In the case of genetic therapy, those electing to use genetic technologies may also be operating under this *a priori* assumption as well.

Third, Philip Peters makes the point that resources are allocated in a way to maximize health care outcomes and maintain nonpartisan neutrality. Peters cautions, however, that "[a]ny health care allocation scheme which attempts to maximize health care outcomes by giving priority to the most effective treatments has the potential to disfavor disabled patients and others, such as the elderly and the frail, whose quality of life is most impaired or whose conditions are most resistant to cure" (1995, 492). Unlike the sort of bias stated above, which assumes a kind of "normality" for health standards, this second assumption concerns the type of disease one may have. If the condition is considered resistant to cure, then it may be best to eliminate that condition using genetic technologies.

What is problematic is that some conditions, such as Down syndrome and Williams syndrome, do not have a cure, but it is still possible for the individual to have a good quality of life. The result is that those electing to use genetic technologies to eliminate disabilities may either, in the first case, hold to a biased standard of normal embodiment or, in second case, may have made generalizations when thinking about incurable health conditions. These points suggest the grounds to conclude that the transhumanists, in their vision of the future, have generally had a "lack of imagination as designers" (Kamm 2005, 13). Further, if we grant that people with disabilities have epistemic privilege about their overall quality of life, then their perspectives should be given respect in the enhancement debate.

To return to the question I asked at the beginning of this section: is there a difference between a life not worth living and a life not worth paying for? If one makes the argument that disability is an unfair cost burden on society, then there is no difference, because the conclusion to eliminate or reduce these lives is the same. However, if one like Singer, turns to arguments about pleasure and pain to indicate a life not worth living, then there is a distinction that could be drawn. And that is where I turn to next.

WHEN PAIN AND SUFFERING ARE NOT THE SAME

In 1999, the disability rights group *Not Dead Yet* protested Singer's appointment at Princeton University by blocking entrance to the university's main building. In protest, they "chanted and held placards that no one should have to prove their personhood" (Lanier 2020, 260). And yet, Singer remains at Princeton. He and other philosophers argue that happiness, especially pleasure according to the Utilitarian calculus, makes life worth living. Pleasure is an indication of a good quality of life. And pain is an indication of a life worse off.

Yet, I think there is a difference between a painful life and a life filled with suffering. As Kyle has said, he does spend every day in pain. And some days are better than others. Pain and discomfort make us aware of the body (Wendell 1996, 170). And Kyle maintains his yoga routine as a way to respond to his body, that is, to listen to what his body is telling him.

While Kyle's health condition has brought him chronic pain, it has been the community around him that has either enabled his flourishing or been the source of his sorrow. If there is one thing I have learned in life, and I think Kyle would agree, it is that both love and hate can bring suffering into our lives. It is a facet of human existence: we are vulnerable bodies, and we are vulnerable to each other. Both give us the capacity to suffer.

Pain is not always suffering because there is a difference in the source. I think philosophers have confused pain, which can be physical or circumstantial, with the existential dimension of suffering. We can target sources of pain which originate in the body and alleviate them, but we cannot target a source of suffering because it is a dimension of lived experience (Bueno-Gómez 2017).

When the face of suffering comes into our lives, we ask the perennial and inescapable question, *why me*? (van Hooft 1998, 19; cf. Lorde 1988, 110). Our cry is a quest for meaning, to find our place in this world. And yet, silence is the only response. Receiving no answer, we wrestle with our affliction on our own. Damaging our selfhood, crushing our spirit, our agony throws us into despair. Suffering humiliation and degradation, we succumb to alienation in relation to others. Desperately, we search for the answer to *why* this is happening to us.

Pain, whether acute or chronic, is experienced differently. It does not cause us to ask *why*. Instead, it directs our attention to *where*. For example, in acute and nonchronic pain, our experience is episodic, it demands our attention, and it constricts our perception of space to the body and of time to the here-and-now. The goal is to get rid of it. And this goal becomes the focus of our intentions and actions (Wendell 1996, 170–71). Nevertheless, we can provide an answer to the question, *where* does it hurt?

Unlike pain, the answer to the question that suffering poses must come from inside ourselves: we must provide a reason for *why* this has happened to *us*. For some, we interpret our suffering as part of a higher plan. That is, "our bodies might suffer maladies, we might suffer pain, our zest for life might be lost, our relationships shattered, our projects failures" but it is still part of an ultimate good (van Hooft 1998, 14). For others, this silence opens the space for love and faith (cf. Weil 1977). Others search for a different answer and, by contrast, embrace a tragic sense of life. They turn inward in hopes of achieving a state invulnerable to life's whims (May 2017). And finally, there are those who see our suffering as a basis for compassion and our communion with others (cf. Levinas 1998). It becomes a way to ground our ethical obligations to one another.

With suffering, there is hope for redemption and self-transcendence. This hope is common in many of the world religions such as Hinduism, Buddhism, and Christianity (van Hooft 1998, 14–15). As a way to ground our ethical obligations to each other or life purpose, suffering becomes integrated into part of a larger moral project. In this way, it becomes an intrinsic part of our moral lives. To suffer for something, then, is to give up something we want for the sake of something more worthy or noble (van Hooft 1998, 15). It is our commitment to this higher project that will make the wrongs we have endured right in the end.

To suffer *for* something worthier is called sacrifice. It is the meaning we give to our suffering. Lanier points out that grief and joy go hand in hand (2020, 262). The point of life is neither to avoid all suffering nor be happy all the time. According to Lanier, the "belief in the virtue of the 'happy' and suffering-free life sterilizes and shrinks us, minimizing what makes us most beautiful human: our tenderness, our vulnerability, the profundity of our capacity for heartache, the risk of which deliver us into immense joy" (ibid., 275–76).

The reason we fear suffering is that it represents a crisis of faith: faith in our ideals, in our belief in a just world, and in our belief in our better nature. There is a general belief that suffering should not happen, because it is inherently negative. For some, death is the only relief we can find for a world of suffering, uncertainty, and calamity. The face of vulnerability that is suffering can come to us in illness, facing death, and experiencing misfortune. It can visit us in old age as isolation and the loss of loved ones. It makes a home in our hearts when they are filled with grief. We fear suffering because it is meaningless: there is no reason for it, no explanation of why it is happening. There is no external answer for *why* it is happening *to me.*

The suffering we imagine in a life not worth living is one in which suffering is not part of an overarching life plan or path to redemption and self-transcendence. It is the inevitable result of bad luck and, as a state of being, it is only to be borne until we die. Suffering without meaning, then, is the ultimate test of our existential faith and commitment to justice. As a face of vulnerability, then, suffering tests our belief in a conception of the good life. And this is why we fear it. In this vein, the challenge in our post-modern world is to forge an authentic acceptance of a suffering that is not made meaningful (van Hooft 1998, 19). To accept that there is a state of affliction, of anguish, of turmoil for which there is no larger story of redemption and transcendence.

The reason I do not find the philosophers' arguments concerning a life not worth living convincing is because they have either reduced suffering to pain in a quantifiable manner or abstracted the concept of suffering from any meaningful context. There is no reason that we suffer. We cannot quantify our suffering. And we cannot retreat from it into abstraction. The experience of suffering is particular, personal, and existential. It makes us question our conception of a good life and a just world. And we cannot control whether or not we experience it in our lifetime because it is a facet of the human condition. All we can do is respond to it in our own diverse ways.

The lived experience of suffering is losing our place in this world with no hope of finding it again. It is a life without redemption. And it is when we feel that we do not or no longer belong here. It is a life lived in alienation to others, to objects, and to ourselves. I believe that it is our friends and loved ones who give us hope again. They are the ones who help us, solitary travelers, find a home and a community. And it is in community with others that we can find solidarity.

WELL-BEING AND THE VIRTUE OF MUTUAL RECOGNITION

Finding my "home" in this world is the gift Kyle gave me. His wisdom and tender way of viewing our vulnerable bodies in this world helped me realize that it is up to us, not science, to shape the society for which we hope. It is up to us, not science, to give our children a future better than ours.

When my husband and I married, Kyle was in our bridal party. At the wedding reception, Kyle became the center of attention because his focus was on the joy he could bring to others. He brought joy with a simple prop, a hat, which he passed around to those attending. If you received the hat, then you had to lead others in dance to the songs the disc jockey played. All the diverse bodies of family and friends joined the dance floor in celebration. Filled with joy, with love, and with laughter, our celebration was one of hope and one of co-belonging.

Kyle taught us the beauty and joy that comes from belonging, from respect, and from the recognition of embodied difference. He taught me what is essential for a good and meaningful life. And that a good life begins in solidarity with others.

In the philosophical debates concerning quality of life, philosophers have often turned to conceptions of welfare and distributive justice. They have argued for their position in terms of fairness or pleasure. In response, I think that not taking into account the perspectives of those who live with a disability commits an epistemic injustice. To rectify this injustice, then, the virtue of mutual recognition is needed. And it has two features.

Two Features of Mutual Recognition

In solidarity, I think that valuing embodied difference is the foundation for having a good quality of life. To value our embodied difference requires, as a first feature, mutual respect. I understand mutual respect much like Silvers and Francis (2005), who propose trust and the value of vulnerability as central for moral interaction and mutual respect (cf. Silvers 1995). Mutual respect extends to group and cultural identities because one's social identity has an effect on one's well-being. It recognizes the axes of power that operate within cultural and social practices and advocates for the mutual respect for individual differences (Moser 2006).[12]

Second, Nancy Fraser (2003) and other critics have argued that social identity cannot be adequately measured by individual advantage within the redistribution framework. People with disabilities are often stigmatized (Wendell 1989). In a line, the aim of solidarity is not only mutual respect, but also interdependent recognition.

Mutual respect and recognition within a society are needed to rectify oppression and stigma; this is because the distributive and utilitarian frameworks of justice are incomplete (Fraser 2016). As Fraser suggests, what is needed is a political dimension in this discussion concerning quality of life:

> the political in this sense furnishes the stage on which struggles over distribution and recognition are played out. Establishing criteria of social belonging, and thus determining who counts as a member, the political dimension of justice specifies the reach of those other dimensions: it tells us who is included in, and who excluded from, the circle of those entitled to a just distribution and reciprocal recognition. Establishing decision rules, the political dimension likewise sets the procedures for staging and resolving contests in both the economic and the cultural dimensions: it tells us not only who can make claims for redistribution and recognition, but also how such claims are to be mooted and adjudicated. (2016, 17)

The epistemic injustice done to people with disabilities in the debate concerning quality of life needs to be rectified. That is done through respect, recognition, and listening in the political dimension. We must ask, have those affected by these decisions been given equal voice? Or have those who are affected been excluded from or silenced in the conversation?

With regard to the use of genetic technologies within society, the aim must be to increase societal well-being from the "outside in" rather than the "inside out." Without changing present societal injustices such as oppression and discrimination, genetic enhancement will only exacerbate those societal inequalities. As Fraser argues, "[o]vercoming injustice means dismantling institutionalized obstacles that prevent some people from participating on par with others, as full partners in social interaction" (2016, 16). An account of social justice that ignores interdependence, solidarity, and recognition is thus incomplete.

We have to remember that disability is a relation between individuals and the environment, and not a capacity individuals lack. And it is the structure of capitalism which disables us and alienates us in our present society. For mutual recognition, the concern is not that people gain respect from others but rather that people respect one another and respect their bodily differences. I argue that the solidarity view recognizes the fact that justice within a future enhanced society must rectify discrimination and oppression. By understanding that justice includes not only equal opportunity but also equal respect and recognition of bodily difference, then, a plan of action and policy can emerge regarding genetic technologies for our future global society.

Application to Facing Suffering

Disability scholars and activists have critiqued genetic technologies because philosophers view disability as an intrinsically bad state rather than as a mere-difference in three ways.[13] First, philosophers who use quality of life measurements invoke a kind of biomedical narrative, which considers impairments as a series of physiological flaws and views medical crises that demand aid "through technology or other allopathic measures" (Garland-Thomson 2005, 1567). Second, they suppose disability to be intrinsically bad by using a statistical understanding of "normal quality of life" to distribute medical resources rather than considering the intersection between disability and class for individual well-being is problematic. Third, by using measurements like QALY, they create a kind of anxiety or hysteria about disability, which reinforces stigma about living with a disability.[14] Thus, these philosophers promote oppression rather than inclusion and co-belonging.[15]

In response to the face of suffering, I think it is necessary to redefine well-being in a holistic perspective, in which time bends to the joy a space can create. In "The Beauty of Spaces Created for and by Disabled People," s.e. smith provides an example of the mutual respect and interdependent recognition found in a theater:

> I look out across the crowd, to the wheelchair and scooter users at the front of the raked seating, the ASL interpreter in crisp black next to the stage. Canes dangle from seat backs and a gilded prosthetic leg gleams under the safety lights. A blind woman in the row below me turns a tiny model of the stage over in her hands, tracing her fingers along with it in time to the audio transcription. (Wong 2020, 272)

The theatrical setting smith describes is a public crip space specifically designed for those who have been marginalized. Mutual respect and interdependent recognition begin with a vivid sense of belonging, with shared experiences, and with a "deep *rightness* that comes from not having to explain or justify your existence" (ibid.). I think that this example of crip space provides a model for how we should view well-being and what a good life should include.

When we consider what could be included in a conception of a good life, I think that a holistic perspective provides some insight (cf. Armstrong 2011; Silberman 2015). When we consider quality of life from a holistic perspective, "it goes beyond activities of daily living and disease categories because it directs attention to the more complete social, psychological and spiritual being" (Albrecht and Devlieger 1999, 979). A number of factors can influence a person's health and well-being, such as the distribution of educational resources:

> For example, Ross and Willigen (1997) found that education improves well-being because it increases access to nonalienated paid work and economic resources that increase the sense of control over life, as well as access to stable social relationships. Education therefore seems to be a component that potentially impacts balance. Likewise, strong social support networks and community ties offer promise in buffering people with disabilities from stress, helping them maintain a balance and anchoring them in the daily activities of the community. On the other hand, pain and isolation are found to be negatively associated with quality of life. (ibid.; cf. Cohen and Wills 1985; Cohen et al. 2001)

If having a strong social support system, ties to one's community, employment and educational resources have a positive impact on one's health and well-being whereas pain and isolation have a negative impact on one's health and well-being, then it would make sense to allocate more resources to people living with disabilities rather than allocating fewer resources.

Should it not be our aim to use genetic technologies to enhance the human experience? If this should be our aim, then our goal should be to increase the access to resources which improve well-being in the present rather than hoping for a better future. Improving someone's linguistic ability or strength, for example, seems to forget the larger picture of a conception of a good life. Similarly, enhancing someone's musical or mathematical abilities will not open more opportunities if one's society has limited economic and educational pathways for the development of those gifts. We must be careful not to "submerge the individual into [the] disability" when projecting into the future (Stein 2006). Furthermore, concerning the use of genetic enhancement technologies, if those living with a disability have a kind of epistemic privilege about their human experience, then philosophers would be wise to listen (Lindemann 2001).

My experience living with a disability and my friendship with Kyle have taught me that there is beauty in the complexity of human embodiment and human experience. These complexities are interconnected with social and cultural meanings. Moreover, these complexities have helped me understand that it is up to us to change our societal practices and to question our beliefs about quality of life and well-being. In other words, the recognition of our own bodily difference and the bodily difference of others provides a dynamic foundation to reconceive of what a community of mutual co-belonging could be.

CONCLUDING THOUGHTS: WHAT DO WE OWE OUR (FUTURE) CHILDREN?

At this point it may be appropriate to step back and consider whether the initial question concerning the use of genetic technologies is on target. Is genetic enhancement an expression of dissatisfaction with a low quality of life or is genetic enhancement a chance to improve ourselves? I argue that this line of reasoning is misguided. Instead, we should ask: is not genetic enhancement's real target, assuming no nefarious aims, the chance to improve the human experience? If so, then to improve human experience for the future, we first need to make changes for human experience in the present.[16]

Second, we need to reconsider the concept of a good "quality of life." The central issue could be that we have mistaken the means of enhancement with the aims. The focus on *what to improve* has been amiss in many areas: various skills such as memory improvement, greater stamina and mental focus, appreciation for music and art, mathematical reasoning and language faculties. While it might be nice to be able to have a better memory and to pass an elementary French class with ease, do these aims not confuse the original goal of human enhancement, i.e., to have a better quality of life?

At the beginning of this work, I asked "what do we owe our (future) children?" I wanted to know how capitalism distorts what we owe each other and how it distorts what we owe our (future) children. This book has been an attempt to answer these concerns. The global conditions of our present society give us concern as scientific developments outpace our normative commitments. The task before us is not a reconceptualization of what is right, but of what is good.

The reason we fear the five faces of vulnerability is because they lead us to question a central belief that we hold: the "universe will not place before us a task we cannot accomplish. It will not, or at least need not, push us beyond our limits" (May 2017, 88). This is because we want to believe that a good life is available to us all, and we only have to figure out how to live it. And it is genetic technologies that promise us that this good life is within our control. It is within reach if we only control our bodies.

And yet, in reality, the five faces indicate not only that much of life is out of our control, but also that the world is not as just as we hope. Controlling our bodies does change the future injustices within a society. Only by changing societal structures can the availability of the good life be realized.

To reconceive of the good requires humility (cf. Coady 2011; Solomon 2015). The virtues of solidarity promote relational well-being, which is enriched by inclusion, interdependence, authenticity, empowerment, openness, trust, diversity, and mutual respect. What my experience with fertility treatments and the adoption process taught me, was that with or without enhancement technologies, one cannot aim to give one's child a good life if one is not first a good parent.

Our children's well-being is relational—it is relational to us and to the society we shape now for their future. Further, the society we mold must provide enough goods and equip its children with opportunities to flourish. This requires an acknowledgement of the systemic features as well as the medical features that reduce health and well-being. The aim of enhancement technologies should be to improve the human experience, but that improvement depends upon us recognizing the beauty and value of bodily difference and variation.

If we do not know love, how can we teach our future children what love is? If we cannot find beauty, how can we teach our future children to search for it? If we cannot accept our own vulnerability, how can we teach our future children to accept their own? As a society, we have an obligation to care for others regardless of biological relation (Liao 2009). Our care for others has led to policy changes such as human rights and civil liberties and to institutional changes such as adoption and educational opportunities. As humans, we are fallible, and we are vulnerable. Enhancement technologies will not be able to change that. Without reflecting on our present moral faults, how can we shape the future for which we hope? To shape the future we hope for our children requires that we change our practices, institutions, policies and environment. It requires that we act in solidarity.

NOTES

1. Philosophers have provided a variety of reasons against the selection of an embryo with a trait for a disability. As a consequentialist, Laura Purdy has argued that she "cannot ignore the probable difficulties that await children with special problems. It seems to me that only the truly rich can secure the well-being of those with the most serious problems. Given the costs and other difficulties of guaranteeing good care, even very well-to-do individuals might well wonder whether their offspring will get the care they need after their own deaths" (1995, 304). Carl H. Coleman (2002), by contrast, uses a comparative framework and has argued that we should consider how the risks and benefits of requested treatment compare to other reproductive and

parenting options. Third, Janet Malek claims that there is no moral difference between using "reproductive genetic technologies to prevent disability in a future child" from protecting oneself from "becoming disabled . . . In each case, an individual is faced with making a choice between a life with a disability and one without. The choice to avoid creating a life with disability may reflect a negative view of the disability itself, but not of persons who have it" (2010, 221). Finally, Buchanan et al. take a welfarist approach and have argued that we "devalue disabilities because we value the opportunities and welfare of the people who have them . . . there is nothing irrational, motivationally incoherent, or disingenuous in saying that we devalue the disabilities and wish to reduce their incidence while valuing existing persons with disabilities, and that we value them the same as those who do not have disabilities" (2000, 278). See also Davis (1997, 2010) against Feinberg's (1980) *right to an open future* argument.

2. This quote was taken from Kafer, p. 52; Quote is originally from Ashley's Mom and Dad, The "Ashley Treatment" March 25, 2007.

3. See also Amundson 2000; Hall 2017; Barnes 2018.

4. The empirical literature provides support for the evidence of stigma and discrimination. In the case of sexual activities and the need for reproductive healthcare services, Xanthe Hunt et al. measured the biases nondisabled individuals in South Wales held concerning people with disabilities: "That is, societal fear regarding the "abnormal" sexuality of people whose bodies are not typical underlies the imperative to desexualise people with physical disabilities, for fear that their sexuality—if it were to be acknowledged—would be somehow monstrous and uncomfortably different, and their offspring somehow genetically or otherwise tainted. The sexuality, and offspring, of people with physical disabilities cannot be normal, and people with physical disabilities, therefore, must not be sexual and must refrain from child-bearing. Such social representations of people with physical disabilities as desexualised and unsuitable or unlikely parents might very well underlie the significant effects for the differences between non-disabled respondents' estimation concerning the sexual rights of people with physical disabilities and the population without disability" (2017, 73). The study performed by Hunt et al. provided some of the first empirical evidence supporting the claims of people with disabilities about bias from nondisabled people concerning their romantic interests and child-rearing capabilities.

5. The phrase "global economy of debility" is referenced in Kola´rˇova, K. and Wiedlack, M. K. 2016.

6. The measurement DALY includes "[y]ears lost from premature mortality . . . estimated with respect to a standard expectation of life at each age. Years lived with disability are translated into an equivalent time loss by using a set of weights which reflect reduction in functional capacity, with higher weights corresponding to a greater reduction" (Anand and Hanson 1997).

7. The use of DALYs, however, often prevents people with disabilities from life-saving interventions. The assumption at play in DALYs is that lives are more valuable if they "contain" more utility, and that utility includes lifespan (Brock 2005). This assumption is false because it does not include the fact that many people with disabilities live rewarding lives (Barnes 2009a; Asch and Wasserman 2010). Using

allocation measures, such as QALYs and DALYs, to direct the distribution of medical resources, then, is unfair to people with disabilities.

8. In "Resource Allocation," Albert R. Jonsen and Kelly A. Edwards list a series of questions which usually guide the allocation of resources: What rules guide rationing decisions? Are there ethical criteria for making triage decisions? Can I make allocation decisions based on judgments about "quality of life"? What about "macro-allocation" concerns? Can we ethically qualify a "right to health care"?

9. The issue at stake concerns whether the burden of living with a disability results from one's social environment and a lack of resources, such as access to education, adequate healthcare, and employment opportunities, or from one's health condition. Exclusively, posed as a question, what is the source of a person with disabilities' low quality of life? Is it the disability itself or the lack of resources? The debate exists because disability is not just any human difference like gender or ethnicity, or any social disadvantage, like poverty or marginalization. First, disability, as a difference, is conceptually linked to an individual's state of health: the combining factors of the physical, the social, and the attitudinal environment affect the functioning and flourishing of a person living with a disability. Second, just as disability is a difference, some also view it as a disadvantage concerning one's health state: this individual needs additional or supplemental resources in order to have a good quality of life, which is unlike race and gender. Because of this second factor, disability is often stigmatized when the allocation of medical resources arises for consideration.

10. Tremain (2017, 172) notes that these are often capability theorists who assume perfectionist conceptions of the good.

11. Some disability scholars point out that the adaptive preference argument is guilty of question begging (Barnes 2009a, 2009b).

12. Sometimes, medical counsel ignores the ways "social structures shape how people experience the possibilities of forming a family" (Shanley and Asch 2009, 852). Instead, medical counsel should exhibit mutual respect and be open to different ways of understanding family formation. Inhibiting parents with disabilities from choosing to have a child with the same social identity creates a society in which it is harder for everyone to flourish.

13. For example, Anand and Wailoo argue that moral frameworks in addition to utilitarianism, such as egalitarianism and concern for human rights, need to be included for the allocation of medical resources (2000, 543). Anand and Wailoo suggest that "QALY maximization has limited appeal because it is a consequentialist social choice rule which has little to say about the multiplicity of rights" for those affected by these debates (ibid., 544).

14. For examples, see Mary Crossley (2000, 63) for the challenges in law concerning the Americans with Disabilities Act and medical treatment for people with disabilities. See also Daniels et al. (2000).

15. Some deontologists have entered this debate. For a deontological perspective, see Jonsen (1986).

16. In contrast to traditional theories of justice, which usually take competition for resources or power to be essential for social and political relationships, one should consider inclusiveness and interdependence and recognize how power, specifically

biopower, can operate within a society (cf. Hill 2017). Further, opportunities should be created for people with disabilities in genetic enhancement discussions.

Bibliography

Aalfs, C. M., E. M. A. Smets, and N. J. Leschot. 2007. "Genetic Counselling for Familial Conditions during Pregnancy: A Review of the Literature Published during the Years 1989 – 2004." *Community Genetics* 10 (3): 159–68. https://doi.org/10.1159/000101757.
Abberly, Paul. 1987. "The Concept of Oppression and the Development of a Social Theory of Disability." *Disability, Handicap & Society* 2 (1): 5–19.
Agar, Nicholas. 2004. *Liberal Eugenics: In Defence of Human Enhancement*. New York: Blackwell.
———. 2007. "Whereto Transhumanism?: The Literature Reaches a Critical Mass." *The Hastings Center Report* 37 (3): 12–17.
———. 2010. *Humanity's End: Why We Should Reject Radical Enhancement*. Cambridge: MIT Press.
———. 2014a. "Moral Bioenhancement and the Utilitarian Catastrophe." *Cambridge Quarterly of Healthcare Ethics* 24 (1): 37–47. https://doi.org/10.1017/S0963180114000280.
———. 2014b. *Truly Human Enhancement: A Philosophical Defense of Limits.*
Cambridge: MIT Press.
Ajf-b. 1983. "Reproductive Technology." In *Off Our Backs: A Women's News Journal*, 13 (5): 5–5.
Albrecht, Gary L., and Patrick J. Devlieger. 1999. "The Disability Paradox: High Quality of Life against All Odds." *Social Science & Medicine* 48: 977–88.
Allen, Amy. 2008. "Power and the Politics of Difference: Oppression, Empowerment, and Transnational Justice." *Hypatia: In Honor of Iris Marion Young: Theorist and Practitioner* 23 (3): 156–72.
Allhoff, Fritz. 2005. "Germ-Line Genetic Enhancement and Rawlsian Primary Goods." *Kennedy Institute of Ethics Journal* 15 (1): 39–56.
Alper, Joseph S. 2002. "Genetic Complexity in Human Disease and Behavior." In *The Double-Edged Helix: Social Implications of Genetics in a Diverse Society*, edited

by Joseph S. Alper, Catherine Ard, Adrienne Asch, Jon Beckwith, Peter Conrad, and Lisa N. Geller, 17–38. Baltimore: Johns Hopkins University Press.

Alper, Joseph S., Catherine Ard, Adrienne Asch, Jon Beckwith, Peter Conrad, and Lisa N. Geller. 2002. "Perspectives on Perspectives." In *The Double-Edged Helix: Social Implications of Genetics in a Diverse Society*, edited by Joseph S. Alper, Catherine Ard, Adrienne Asch, Jon Beckwith, Peter Conrad, and Lisa N. Geller, 1–16. Baltimore: Johns Hopkins University Press. https://doi.org/10.2307/776608.

Amundson, Ron. 1992. "Disability, Handicap and the Environment." *Journal of Social Philosophy* 23 (1): 105–19.

———. 2000. "Against Normal Function." *Studies in History, Philosophy and Biomedical Sciences* 31 (1): 33–53.

Amundson, Ron, and Shari Tresky. 2007. "On a Bioethical Challenge to Disability Rights." *Journal of Medicine and Philosophy* 32 (6): 541–61. https://doi.org/10.1080/03605310701680924.

Anand, Paul, and Allan Wailoo. 2000. "Utilities versus Rights to Publicly Provided Goods: Arguments and Evidence from Health Care Rationing." *Economica* 67: 543–77.

Anand, S, and Karen Hanson. 1997. "Disability-Adjusted Life Years: A Critical Review." *Journal of Health Economics* 16: 685–702.

Anderson, Elizabeth S. 2010. "Segregation and Social Inequality." In *The Imperative of Integration, 1–22*. Princeton: Princeton University Press.

Annas, George J., Lori B. Andrews, and Rosario M. Isasi. 2002. "Protecting the Endangered Human: Toward an International Treaty Prohibiting Cloning and Inheritable Alterations." *American Journal of Law & Medicine* 28 (2–3): 151–78. https://doi.org/10.4324/9781315254517-18.

Armstrong, Thomas. 2011. *The Power of Neurodiversity: Unleashing the Advantages of Your Differently Wired Brain*. Philadelphia: First De Capo Books.

Arneil, Barbara. 2009. "Disability, Self Image, and Modern Political Theory." *Political Theory* 37 (2): 218–42. https://doi.org/10.1177/0090591708329650.

Arnesen, Trude, and Erik Nord. 1999. "The Value of DALY Life: Problems with Ethics and Validity of Disability Adjusted Life Years." *BMJ: British Medical Journal* 319 (7222): 1423–25.

Asch, Adrienne. 1993. "Abused or Neglected Clients - or Abusive or Neglectful Service Systems?" In *Ethical Conflicts in the Management of Home Care*, edited by R.A. Kane and A.L. Caplan, 113–21. New York: Springer.

———. 1999. "Prenatal Diagnosis and Selective Abortion: A Challenge to Practice and Policy." *American Journal of Public Health* 89 (11): 1649–57. https://doi.org/10.2105/AJPH.89.11.1649.

———. 2002. "Disability and Reproductive Rights." *Historical & Multicultural Encyclopedia of Female Reproductive Rights in the United States* 53 (9): 63–67. https://doi.org/10.1017/CBO9781107415324.004.

———. 2003. "Disability Equality and Prenatal Testing: Contradictory or Compatible?" *Florida State University Law Review* 30 (2): 315–42.

Asch, Adrienne, and Michelle Fine. 2009. "Shared Dreams: A Left Perspective on Disability Rights and Reproductive Rights." In *Women with Disabilities: Essays in Psychology, Culture, and Politics*, edited by Adrienne Asch and Michelle Fine, 297–305. Temple: Temple University Press.

Asch, Adrienne, and Gail Getter. 1996. "Feminism, Bioethics, and Genetics." In *Feminism & Bioethics: Beyond Reproduction*, edited by Susan M. Wolf, 318–50. Oxford: Oxford University Press.

Asch, Adrienne, Lawrence O. Gostin, and Diann M. Johnson. 2003. "Respecting Persons with Disabilities and Preventing Disability: Is There a Conflict?" In *The Human Rights of Persons with Intellectual Disabilities: Different but Equal*, edited by Stanley S. Herr, Lawrence O. Gostin, and Harold Hongju Koh, 319–46. New York: Oxford University Press. https://doi.org/10.5860/choice.41-6777.

Asch, Adrienne and David Wasserman. 2005. "Where Is the Sin in Synecdoche: Prenatal Testing and the Parent-Child Relationship." In *The Quality of Life and Human Difference: Genetic Testing, Health Care, and Disability*. Edited by David Wasserman, Jerome Bickenbach, and Robert Wachbroit, 172–216. New York: Cambridge University Press.

———. 2010. "Making Embryos Healthy or Making Healthy Embryos: How Much of a Difference between Prenatal Treatment and Selection?" In *The "Healthy" Embryo: Social, Biomedical, Legal and Philosophical Perspectives*, edited by J. Nisker, Françoise Baylis, I. Karpin, Carole McLeod, and R. Mykitiuk, 201–19. New York: Cambridge University Press.

Aschbrenner, Kelly A., Jan S. Greenberg, Susan M. Allen, and Marsha Mailick Seltzer. 2010. "Subjective Burden and Personal Gains Among Older Parents of Adults with Serious Mental Illness." *Psychiatric Services* 61 (6): 605–11. https://doi.org/10.1176/ps.2010.61.6.605.

Ashley's Mom and Dad. "The 'Ashley Treatment': Towards a Better Quality of Life for 'Pillow Angels'." Pillow Angels, last modified March 25, 2007. http://pillowangel.org/Ashley%20Treatment.pdf.

Azevedo, Marco Antonio. 2016. "The Misfortunes of Moral Enhancement." *Journal of Medicine and Philosophy* 41 (5): 461–79. https://doi.org/10.1093/jmp/jhw016.

Badenoch, Bonnie. 2008. *Being a Brain-Wise Therapist: A Practical Guide to Interpersonal Neurobiology*. New York: W.W. Norton & Company.

Baier, Annette. 1986. "Trust and Antitrust." *Ethics* 96 (2): 231–60.

———. 1987. "The Need for More Than Justice." *Canadian Journal of Philosophy* 13: 41–56.

Baltussen, Rob, and Louis Niessen. 2006. "Priority Setting of Health Interventions: The Need for Multi-Criteria Decision Analysis." *Cost Effectiveness and Resource Allocation* 4 (14): 1–9. https://doi.org/10.1186/1478-7547-4-14.

Baril, Alexandre. 2016. "Doctor, am I an Anglophone trapped in a Francophone body. An intersextional analysis of 'trans-frip-t time' in ableist, cisnormative, anglonormative societies." *Journal of Literary & Cultural Disability Studies* 10 (2): 155–72. https://doi.org/10.3828/jlcds.2016.14

Barnes, Elizabeth. 2009a. "Disability, Minority, and Difference." *Journal of Applied Philosophy* 26 (4): 337–55. https://doi.org/10.1111/j.1468-5930.2009.00443.x.

———. 2009b. "Disability and Adaptive Preference." *Philosophical Perspectives* 23 (Ethics): 1–22.

———. 2014. "Valuing Disability, Causing Disability." *Ethics* 125 (1): 88–113. https://doi.org/10.1086/677021.

———. 2018. *The Minority Body: A Theory of Disability*. New York: Oxford University Press.

Bartky, Sandra Lee. 2005. "On Psychological Oppression." In *Feminist Theory: A Philosophical Anthology*, edited by Anne E. Cudd and Robin O. Andreasen, 105–14. New York: Blackwell.

Baylis, Françoise, and Jason Scott Robert. 2004. "The Inevitability of Genetic Enhancement Technologies." *Bioethics* 18 (1): 17–27.

Becker, Lawrence C. 2000. "The Good of Agency." In *Americans with Disabilities: Exploring Implications of the Law for Individuals and Institutions*, edited by Leslie Pickering Francis and Anita Silvers, 54–63. New York: Routledge.

———. 2005. "Reciprocity, Justice, and Disability." *Ethics* 116 (1): 9–39.

———. 2012. "Concepts and Conceptions: Basic Justice and Habilitation." In *Habilitation, Health, and Agency: A Framework for Basic Justice*, 1–18. Oxford: Oxford University Press. https://doi.org/10.1093/acprof.

Beckwith, Jon. 2002. "Geneticists in Society, Society in Genetics." In *The Double-Edged Helix: Social Implications of Genetics in a Diverse Society*, edited by Joseph S. Alper, Catherine Ard, Adrienne Asch, Jon Beckwith, Peter Conrad, and Lisa N. Geller, 39–56. Baltimore: Johns Hopkins University Press.

Bentham, Jeremy. 1789. *An Introduction to the Principles of Morals and Legislation*, Chapter 3. https://www.econlib.org/library/Bentham/bnthPML.html?chapter_num=4#book-reader.

Bickenbach, Jerome E. 1993. "Introduction and Overview." In *Physical Disability and Social Policy*. Toronto: University of Toronto Press. https://doi.org/10.2307/3552057.

Bijma, Hilmar H., Agnes van der Heide, and Hajo I. J. Wildschut. 2008. "Decision-Making after Ultrasound Diagnosis of Fetal Abnormality." *Reproductive Health Matters* 16 (31): 82–89.

Bognar, Greg. 2011. "Impartiality and Disability Discrimination." *Kennedy Institute of Ethics Journal* 21 (1): 1–23.

Bolaki, Stella. 2019. "The Cancer Journals (1980)." *In Disability Experiences: Memoirs, Autobiographies, and Other Personal Narratives*, edited by G. Thomas Couser and Susannah B. Mintz, 110-14. New York: Macmillan.

Boorse, Christopher. 1987. "Concepts of Health." In *Health Care Ethics: An Introduction*, edited by Donald VanDeVeer and Tom Regan, 359–93. Temple: Temple University Press.

Bordo, Susan. 1993. *Unbearable Weight: Feminism, Western Culture, and the Body*, Berkeley: University of California Press.

Bostrom, Nick. 2003. "Human Genetic Enhancements: A Transhumanist Perspective." *The Journal of Value Inquiry* 37: 493–506.

———. 2004. "The Future of Human Evolution." In *Death and Anti-Death: Two Hundred Years After Kant, Fifty Years After Turing*, edited by Charles Tandy, 339–371. Palo Alto: Ria University Press.

———. 2005a. "A History of Transhumanist Thought." *Journal of Evolution and Technology* 14 (1): 1–30.

———. 2005b. "The Fable of the Dragon-Tyrant." *Journal of Medical Ethics* 31 (5): 273–77.

———. 2005c. "Transhumanist Values." *Journal of Philosophical Research* 30: 3–14.

———. 2008a. "Letter from Utopia." *Studies in Ethics, Law, and Technology* 2 (1): 1–7.

———. 2008b. "Why I Want to Be a Posthuman When I Grow Up." In *Medical Enhancement and Posthumanity*, edited by Bert Gordijn and Ruth Chadwick, 107–37. Springer.

Bostrom, Nick, and Toby Ord. 2006. "The Reversal Test: Eliminating Status Quo Bias in Applied Ethics." *Ethics* 116 (4): 656–79.

Bostrom, Nick, and Rebecca Roache. 2008. "Ethical Issues in Human Enhancement." In *New Waves in Applied Ethics*, edited by Jesper Ryberg, Thomas Petersen, and Clark Wolf, 120–52. New York: Palgrave Macmillan.

Bostrom, Nick, and Anders Sandberg. 2008. "The Wisdom of Nature: An Evolutionary Heuristic for Human Enhancement." In *Human Enhancement*, edited by Julian Savulescu and Nick Bostrom, 375–416. Oxford: Oxford University Press.

———. 2009. "Cognitive Enhancement: Methods, Ethics, Regulatory Challenges." *Science and Engineering Ethics* 15 (3): 311–41.

Bostrom, Nick, and Julian Savulescu. 2008. "Human Enhancement Ethics: The State of the Debate." In *Human Enhancement*, edited by Julian Savulescu and Nick Bostrom, 1–22. New York: Oxford University Press.

Boyle, Robert J, and Julian Savulescu. 2001. "Ethics of Using Preimplantation Genetic Diagnosis to Select a Stem Cell Donor for an Existing Person." *BMJ: British Medical Journal* 323 (7323): 1240–43.

Boys, Jos. 2018. "Cripping Spaces? On Dis/Abling Phenomenology: In Architecture." *Log* 42: 55–66.

Braverman, Harry. 1974/1998. *Labor and Monopoly Capital: The Degradation of Work in the Twentieth Century*. New York: Monthly Review Press.

Brock, Dan W. 2003. "Separate Spheres and Indirect Benefits." *Cost Effectiveness and Resource Allocation* 1 (4): 1–12.

———. 2005. "Preventing Genetically Transmitted Disabilities While Respecting Persons with Disabilities." In *Quality of Life and Human Difference: Genetic Testing, Health Care, and Disability*, edited by David Wasserman, Jerome E. Bickenbach, and Robert Wachbroit, 67–100. Cambridge: Cambridge University Press. https://doi.org/10.1017/CBO9780511614590.004.

Bublitz, Jan Christoph, and Reinhard Merkel. 2009. "Autonomy and Authenticity of Enhanced Personality Traits." *Bioethics* 23 (6): 360–74. https://doi.org/10.1007/s10202-008-0060-4.

Buchanan, Allen. 2009. "Moral Status and Human Enhancement." *Philosophy and Public Affairs* 37 (4): 346–81. https://www.jstor.org/stable/40468461.

———. 2011a. *Better Than Human: The Promise and Perils of Enhancing Ourselves*. New York: Oxford University Press.

———. 2011b. *Beyond Humanity? The Ethics of Biomedical Enhancement*. Oxford: Oxford University Press.

Buchanan, Allen, Dan W. Brock, Norman Daniels, and Daniel Wikler. 2000/2009. *From Chance to Choice: Genetics and Justice*. Cambridge: Cambridge University Press.

Buck v. Bell. 1927. 274 U.S. 200. https://supreme.justia.com/cases/federal/us/274/200/

Bueno-Gómez, N. 2017. Conceptualizing Suffering and Pain. *Philos Ethics Humanit Med* **12** (7). https://doi.org/10.1186/s13010-017-0049-5.

Campbell, Stephen M. 2015. "When the Shape of a Life Matters." *Ethical Theory and Moral Practice* 18 (3): 565–75. https://doi.org/10.1007/s10677-014-9540-x.

Cappelen, Alexander W., and Ole Frithjof Norheim. 2006. "Responsibility, Fairness and Rationing in Health Care." *Health Policy* 76: 312–19. https://doi.org/10.1016/j.healthpol.2005.06.013.

Carel, Havi. 2007. "Can I Be Ill and Happy?" *Philosophia* 35 (2): 95–110. https://doi.org/10.1007/s11406-007-9085-5.

———. 2018. *Phenomenology of Illness*. New York: Oxford University Press.

Carlson, Licia. 2001. "Cognitive Ableism and Disability Studies: Feminist Reflections on the History of Mental Retardation." *Hypatia* 16 (4): 124–46. https://doi.org/10.2979/hyp.2001.16.4.124.

———. 2009. "Philosophers of Intellectual Disability: A Taxonomy." *Metaphilosophy* 40 (3–4): 552–66. https://doi.org/10.1002/9781444322781.ch17.

———. 2010. "Who's the Expert? Rethinking Authority in the Face of Intellectual Disability." *Journal of Intellectual Disability Research*. https://doi.org/10.1111/j.1365-2788.2009.01238.x.

———. 2013. "Rethinking Normalcy, Normalization, and Cognitive Disability." *Science and Other Cultures: Issues in Philosophies of Science and Technology*, 154–71. https://doi.org/10.4324/9781315881010.

Carlson, Licia, and Eva Feder Kittay. 2010. "Introduction: Rethinking Philosophical Presumptions in Light of Cognitive Disability." In *Cognitive Disability and Its Challenge to Moral Philosophy*. https://doi.org/10.1002/9781444322781.ch1.

CBC Docs. 2019. "22 with terminal cancer: Canadian gymnast grapples with a life-changing diagnosis | Before & After." YouTube. October 13, 2019. https://www.youtube.com/watch?v=czlMz5IhUww

Chan, Sarah, and John Harris. 2007. "In Support of Human Enhancement." *Studies in Ethics, Law, and Technology* 1 (1).

———. 2011. "Moral Enhancement and Pro-Social Behaviour." *Journal of Medical Ethics* 37 (3): 130–31.

Clare, Eli. 2017. *Brilliant Imperfection: Grappling with Cure*. Durham: Duke University Press.

Coady, C. A. J. 2011. "Playing God." In *Human Enhancement*, edited by Julian Savulescu and Nick Bostrom, 155–80. Oxford: Oxford University Press. https://doi.org/10.1007/978-0-387-69039-1_2.

Coady, David. 2012. "Two Concepts of Epistemic Injustice." *Episteme* 7 (2): 101–13. https://doi.org/10.3366/E1742360010000845.

Cohen, G.A. 2000. *Karl Marx's Theory of History: A Defence*. Princeton: Princeton University Press.
Cohen, Sheldon, Benjamin H. Gottlieb, and Lynn G. Underwood. 2001. "Social Relationships and Health: Challenges for Measurement and Intervention." *Advances in Mind-Body Medicine* 17: 129–41.
Cohen, Sheldon, and Thomas Ashby Wills. 1985. "Stress, Social Support, and the Buffering Hypothesis." *Psychological Bulletin* 98 (2): 310–57. https://doi.org/10.1037/0033-2909.98.2.310.
Cokley, Rebecca. 2020. "The Antiabortion Bill You Aren't Hearing About. In *Disability Visibility: First-Person Stories from the Twenty-First Century*, edited by Alice Wong, 159–64. New York: Vintage Books.
Coleman, Carl H. 2002. "Conceiving Harm: Disability Discrimination in Assisted Reproductive Technologies." *UCLA Law Review* 50 (1): 17–68. https://doi.org/10.1017/CBO9781107415324.004.
Collings, Susan, and Gwynnyth Llewellyn. 2012. "Children of Parents with Intellectual Disability: Facing Poor Outcomes or Faring Okay?" *Journal of Intellectual and Developmental Disability* 37 (1): 65–82. https://doi.org/10.3109/13668250.2011.648610.
Connor, David, Beth A. Ferri, and Subini A. Annamma, eds. 2016. *DisCrit: Disability Studies and Critical Race Theory in Education*. New York: Teachers College Press.
Conrad, Peter. 1997. "Public Eyes and Private Genes: Historical Frames, News Constructions, and Social Problems." *Social Problems* 44 (2): 139–54. https://doi.org/10.2307/3096939.
———. 2002. "Genetics and Behavior in the News: Dilemmas of a Rising Paradigm." In *The Double-Edged Helix: Social Implications of Genetics in a Diverse Society*, edited by Joseph S. Alper, Catherine Ard, Adrienne Asch, Jon Beckwith, Peter Conrad, and Lisa N. Geller, 58–79. Baltimore: Johns Hopkins University Press.
Cookson, Richard, and Paul Dolan. 1999. "Public Views on Health Care Rationing: A Group Discussion Study." *Health Policy* 49: 63–74.
Corker, Mairian. 2001. "Sensing Disability." *Hypatia* 16 (4): 34–52.
Crook, Paul. 2008. "The New Eugenics? The Ethics of Bio-Technology." *Australian Journal of Politics and History* 54 (1): 135–43. https://doi.org/10.1080/09687599.2015.1045353.
Crossley, Mary A. 2000. "Becoming Visible: The ADA's Impact on Health Care for Persons with Disabilities." *Alabama Law Review* 52 (1): 51–90.
Cudd, Ann. 2006. *Analyzing Oppression*. New York: Oxford University Press.
Cureton, Adam and David Wasserman, eds. 2020. *The Oxford Handbook on Philosophy of Disability*. New York: Oxford University Press.
Daniels, Norman.1994. "Four Unsolved Rationing Problems." *The Hastings Center Report* 24 (4): 27.
———. 2000. "Normal Functioning and the Treatment-Enhancement Distinction." *Cambridge Quarterly of Healthcare Ethics* 9: 309–22. https://pdfs.semanticscholar.org/ba5d/015417bb97abc0dc09f36227f6ce1301fdf4.pdf.

Daniels, Norman, J. Bryant, R. A. Castano, O. G. Dantes, K. S. Khan, and S. Pannarunothai. 2000. "Benchmarks of Fairness for Health Care Reform: A Policy Tool for Developing Countries." *World Health Organization* 78 (6): 740–50.

Dar-Nimrod, Ilan, and Steven J. Heine. 2011. "Genetic Essentialism: On the Deceptive Determinism of DNA." *Psychological Bulletin* 137 (5): 800–818. https://doi.org/10.1037/a0021860.Genetic.

Davies, S., and S. Taylor-Alexander. 2019. "Temporal Orders and Y Chromosome Futures: Of Mice, Monkeys, and Men." *Catalyst: Feminism, Theory, Technoscience*, 5 (1): 1–18.

Davis, Dena S. 1997. "Genetic Dilemmas and the Child's Right to an Open Future." *Rutgers Law Journal* 28: 549–92. https://doi.org/10.2307/3527620.

———. 2010. *Genetic Dilemmas: Reproductive Technology, Parental Choices, and Children's Futures*. Oxford: Oxford University Press. https://doi.org/10.1016/j.ajhg.2010.01.004.

Davis, L. J. 2000. "Go to the Margins of the Class: Hate Crimes and Disability." In *Americans with Disabilities: Exploring Implications of the Law for Individuals and Institutions* edited by Leslie P. Francis & Anita Silvers, 331–38. New York: Routledge.

Davis, N. Ann. 2005. "Invisible Disability." *Ethics* 116 (1): 153–213. https://doi.org/10.1086/453151.

de Beauvoir, Simone. 1972/1996. *The Coming of Age*. New York: Warner Books; repr. W.W. Norton.

DeGrazia, David. 2000. "Prozac, Enhancement, and Self-Creation." *The Hastings Center Report* 30 (2): 34–40.

———. 2005. "Enhancement Technologies and Human Identity." *Journal of Medicine and Philosophy* 30 (3): 261–83. https://doi.org/10.1080/03605310590960166.

———. 2008. "Moral Status as a Matter of Degree?" *Southern Journal of Philosophy* 46 (2): 181–98. https://doi.org/10.1111/j.2041-6962.2008.tb00075.x.

———. 2012. "Genetic Enhancement, Post-Persons and Moral Status: A Reply to Buchanan." *Journal of Medical Ethics* 38: 135–39. https://doi.org/10.1136/medethics-2011-100126.

———. 2014. "Moral Enhancement, Freedom, and What We (Should) Value in Moral Behaviour." *Journal of Medical Ethics* 40: 361–68. https://doi.org/10.1136/medethics-2012-101157.

———. 2016. "Ethical Reflections on Genetic Enhancement with the Aim of Enlarging Altruism." *Health Care Analysis* 24 (3): 180–95. https://doi.org/10.1007/s10728-015-0303-1.

De Grey, Aubrey D. N. J. 2003. "The Foreseeability of Real Anti-Aging Medicine: Focusing the Debate." *Experimental Gerontology* 38 (9): 927–34. https://doi.org/10.1016/S0531-5565(03)00155-4.

Diekema, Douglas S. 1990. "Is Taller Really Better? Growth Hormone Therapy in Short Children." *Perspectives in Biology and Medicine* 34 (1): 109–23. https://doi.org/10.1353/pbm.1990.0052.

Dhanda, Rahul K. 2002. *Guiding Icarus: Merging Bioethics with Corporate Interests*. New York: John Wiley and Sons.

Dolnick, E. 1993 (September). "Deafness as Culture." *The Atlantic Monthly*: 37–53.

Douglas, Thomas. 2008. "Moral Enhancement." *Journal of Applied Philosophy* 25 (3): 228–45.

———. 2010. "Genetic Enhancement, Human Nature, and Rights." *Journal of Medicine and Philosophy* 35 (4): 415–28. https://doi.org/10.1093/jmp/jhq034.

———. 2013. "Human Enhancement and Supra-Personal Moral Status." *Philosophical Studies* 162: 473–97. https://doi.org/10.1007/s11098-011-9778-2.

———. 2014a. "Moral Enhancement." In *Enhancing Human Capacities*, edited by Julian Savulescu, Ruud ter Meulen, and Guy Kahane, 465–85. New York: Blackwell. https://doi.org/10.1017/CBO9781107415324.004.

———. 2014b. "The Harms of Enhancement and the Conclusive Reasons View." *Cambridge Quarterly of Healthcare Ethics* 24 (1): 23–36. https://doi.org/10.1017/S0963180114000218.

Dresser, Rebecca. 1995. "Dworkin on Dementia: Elegant Theory, Questionable Policy." *The Hastings Center Report* 25 (6): 32–38. https://doi.org/10.2307/3527839.

Elliott, Carl. 2011. "Enhancement Technologies and the Modern Self." *Journal of Medicine and Philosophy* 36 (4): 364–74. https://doi.org/10.1093/jmp/jhr031.

Epstein, Charles J., and Mitchell S. Golbus. 1977. "Prenatal Diagnosis of Genetic Diseases: Recently Developed Techniques for Examining the Fetus Often Make It Possible to Reduce the Risk of Giving Birth to a Child with a Genetic Disease." *American Scientist* 65 (6): 703–11.

Erevelles, Nirmala. 2011. "Disability as 'Becoming': Notes on the Political Economy of the Flesh." In *Disability and Difference in Global Contexts: Enabling a Transformative Body Politic*, 25–63. New York: Palgrave.

Erevelles, Nirmala, and Xuan Thuy Nguyen. 2016. "Introduction." In *Girlhood Studies, Special Issue "Disability, Girlhood, and Vulnerability in Transnational Contexts"* 9 (1): 3–20. https://doi.org/10.3167/ghs.2016.090102.

Etieyibo, Edwin. 2012. "Genetic Enhancement, Social Justice, and Welfare-Oriented Patterns of Distribution." *Bioethics* 26 (6): 296–304. https://doi.org/10.1111/j.1467-8519.2010.01872.x.

Fabre, C. 2006. "New Technologies, Justice and the Body." In *The Oxford Handbook of Political Theory*, edited by J. S. Dryzek, B. Honig, and A. Phillips, 713-28. New York, NY: Oxford University Press.

Falls, W. A., M. J. D. Miserendino, and M. Davis. 1992. "Extinction of Fear-Potentiated Startle—Blockade by Infusion of an Nmda Antagonist into the Amygdala". *Journal of Neuroscience*, 12 (3): 854–63.

Farrelly, Colin. 2005. "Justice in the Genetically Transformed Society." *Kennedy Institute of Ethics Journal* 15 (1): 91–99.

Feinberg, Joel. 1980. "The Child's Right to an Open Future." In W. Aiken and H. LaFollette (eds.), *Whose Child? Children's Rights, Parental Authority, and State Power*. Totowa, NJ: Rowman and Littlefield.

Ferguson, Ann. 2009. "Feminist Paradigms of Solidarity and Justice." *Philosophical Topics* 37 (2): 161–77.

Ferguson, Philip M. 2001. "Mapping the Family: Disability Studies and the Exploration of Parental Response to Disability." In *Handbook of Disability Studies,* edited by Gary L. Albrecht, Katherine D. Seelman, and Michael Bury, 373–395. Thousand Oaks, CA: Sage Publications.

Flanigan, J. 2017. *Pharmaceutical Freedom: Why Patients Have a Right to Self-Medicate*. Oxford: Oxford University Press.

Foucault, Michel. 1973/1994. *The Birth of the Clinic: An Archaeology of Medical Perception*. New York: Vintage Books.

Francis, Leslie P. 2009. "Understanding Autonomy in Light of Intellectual Disability." In *Disability and Disadvantage*, edited by K. Brownlee and Adam Cureton, 200–215. New York: Oxford University Press. https://doi.org/10.1093/acprof:osobl/9780199234509.003.0008.

Frank, Arthur W. 1995. *The Wounded Storyteller: Body, Illness, and Ethics*. Chicago: University of Chicago Press.

Fraser, Nancy. 2003. *Redistribution or Recognition? A Political-Philosophical Exchange*. Brooklyn: Verso Publishing.

———. 2016. "Capitalism's Crisis of Care." *Dissent* (Fall): 30–38.

Fromm, Erich. 1961. *Marx's Concept of Man*. New York: Frederick Ungar Publishing.

Fricker, Miranda. 2007. *Epistemic Injustice: Power and the Ethics of Knowing*. New York: Oxford University Press.

Fukuyama, Francis. 2002. *Our Posthuman Future: Consequences of the Biotechnology Revolution*. London: Profile Books.

Gabriel, Markus. 2015. *Why the World Does Not Exist*. Trans. by Gregory S. Moss. Malden: Polity Press.

Garde, Jonah I. 2016. "Inclusive Development as Crip (Dys) Topic Promise: Querying Development, Dis/Ability and Human Rights." *Somatechnics, Special Issue: Crip Notes on the Idea of Development* 6 (2): 159–78. https://doi.org/10.3366/soma.2016.0189.

Gardner, William. 1995. "Can Human Genetic Enhancement Be Prohibited?" *Journal of Medicine and Philosophy* 20: 65–84.

Garland-Thomson, Rosemarie. 2001. *Re-Shaping, Re-Thinking, Re-Defining: Feminist Disability Studies*. Washinton, DC: Center for Women Policy Studies.

———. 2005. "Feminist Disability Studies." *Signs* 30 (2) (Winter): 1557–87.

———. 2011"Misfits: A Feminist Materialist Disability Concept." *Hypatia* 26 (June): 591–609.

Geller, Lisa N. 2002. "Current Developments in Genetic Discrimination." In *The Double-Edged Helix: Social Implications of Genetics in a Diverse Society*, edited by Joseph S. Alper, Catherine Ard, Adrienne Asch, Jon Beckwith, Peter Conrad, and Lisa N. Geller, 267–85. Baltimore: Johns Hopkins University Press.

Gems, David. 2011. "Tragedy and Delight: The Ethics of Decelerated Ageing." *Philosophical Transactions: Biological Sciences* 366 (1561): 108–12. https://doi.org/10.1098/rstb.2010.0288.

Gilbert, Daniel. 2006. *Stumbling on Happiness*. New York: Random House.

Giubilini, Alberto, and Sagar Sanyal. 2015. "The Ethics of Human Enhancement." *Philosophy Compass* 10 (4): 233–43.

Glover, Jonathan. 2007. *Choosing Children: Genes, Disability, and Design*. Oxford: Oxford University Press.

Goering, Sara. 2008. "'You Say You're Happy, But . . .': Contested Quality of Life Judgments in Bioethics and Disability Studies." *Journal of Bioethical Inquiry* 5 (2–3): 125–35. https://doi.org/10.1007/s11673-007-9076-z.

Goffman, Erving. 1963. *Stigma: Notes on the Management of Spoiled Identity*. New York: Simon and Schuster.

Goggin, Gerard. 2009. "Disability, Media, and the Politics of Vulnerability." *Asia Pacific Media Educator* 1 (19): 1–13.

Gonick, Marnina. 2006. "Between 'Girl Power' and 'Reviving Ophelia': Constituting the Neoliberal Girl Subject." *NWSA Journal* 18 (2): 1–23.

Gooding, Holly C., Benjamin Wilfond, Karina Boehm, and Barbara Bowles Biesecker. 2002. "Unintended Messages: The Ethics of Teaching Genetic Dilemmas." *The Hastings Center Report* 32 (2): 37–39.

Gordijn, Bert, and Ruth Chadwick. 2008. *Medical Enhancement and Posthumanity*. Edited by Bert Gordijn and Ruth Chadwick. *The International Library of Ethics, Law and Technology* Vol. 2. Springer Netherlands. https://doi.org/10.1007/978-1-4020-8852-0.

Gottlieb, Lori. 2019. "A Dying Woman Tells Us How to Live." *The New York Times*, February 6, 2019. https://www.nytimes.com/2019/02/06/books/review/julie-yip-williams-unwinding-miracle.html.

Guignon, C. 2004. *On Being Authentic*. New York: Routledge.

Gupta, Jyotsna Agnihotri. 2010. "Private and Public Eugenics: Genetic Testing and Screening in India." In *Frameworks of Choice: Predictive and Genetic Testing in Asia*, edited by Margaret Sleeboom-Faulkner, 43–63. Amsterdam: Amsterdam University Press.

Habermas, Jürgen. 2003. *The Future of Human Nature*. Cambridge: Polity Press.

Hahn, H. 1997. "Advertising the Acceptably Employable Image: Disability and Capitalism." In *The Disability Studies Reader*, edited by L.J. Davis, 172–86. London: Routledge Kegan Paul.

Halberstam, Judith. 2005. *In a Queer Time and Place: Transgender Bodies, Subcultural Lives*. New York: New York University Press.

Haliburton, Rachel. 2014. *Autonomy and the Situated Self: A Challenge to Bioethics*. Lanham: Lexington Books.

Hall, Melinda C. 2013. "Reconciling the Disability Critique and Reproductive Liberty: The Case of Negative Genetic Selection." *International Journal of Feminist Approaches to Bioethics* 6 (1): 121–43.

———. 2017. *The Bioethics of Enhancement: Transhumanism, Disability, and Biopolitics*. New York: Lexington Books.

Hamraie, A., and K. Fritsch. 2019. "Crip Technoscience Manifesto." *Catalyst: Feminism, Theory, Technoscience* 5 (1): 1–34.

Harris, John. 1987. "QALYfying the Value of Life." *Journal of Medical Ethics* 13 (3): 117–23.

———. 2011. "Moral Enhancement and Freedom." *Bioethics* 25 (2): 102–11. https://doi.org/10.1111/j.1467-8519.2010.01854.x.

Heidegger, M. 1962. *Being and Time*. Trans. J. Macquarrie and E. Robinson. New York: Harper & Row.

Heredero-Baute, Luis. 2004. "Community-Based Program for the Diagnosis and Prevention of Genetic Disorders in Cuba." *Community Genetics* 7 (2): 130–36. https://doi.org/10.1159/000080783.

Hernstein, R. and Murray, C. 1996. *The Bell Curve: Intelligence and Class Structure in American Life*. New York: Free Press.

Hinson, Katrina and Ben Sword. 2019. "Illness Narratives and Facebook: Living Illness Well." *Humanities* 8 (106). doi:10.3390/h8020106

Hope, Tony. 2011. "Cognitive Therapy and Positive Psychology Combined: A Promising Approach to the Enhancement of Happiness." In *Enhancing Human Capacities*, edited by Julian Savulescu, Ruud ter Meulen, and Guy Kahane, 230–44. New York: Blackwell. https://doi.org/10.1002/9781444393552.ch16.

Horovitz, Dafne Dain Gandelman, Ruben Araújo de Mattos, and Juan Clinton Llerena Jr. 2004. "Medical Genetic Services in the State of Rio de Janeiro, Brazil." *Community Genetics* 7 (2): 111–16. https://doi.org/10.1159/000080779.

Hughes, James J. 1996. "Embracing Change with All Four Arms: A Post-Humanist Defense of Genetic Engineering." *Eubios Journal of Asian and International Bioethics* 6 (June 4): 94–101.

———. 2004. *Citizen Cyborg: Why Democratic Societies Must Respond to The Redesigned Human of the Future.* Cambridge, MA: Westview Press.

———. 2015. "Moral Enhancement Requires Multiple Virtues: Toward a Posthuman Model of Character Development." *Cambridge Quarterly of Healthcare Ethics* 24: 86–95. https://doi.org/10.1017/ S0963180114000334.

Hunt, Xanthe, Mark T. Carew, Stine Hellum Braathen, Leslie Swartz, Mussa Chiwaula, and Poul Rohleder. 2017. "The Sexual and Reproductive Rights and Benefit Derived from Sexual and Reproductive Health Services of People with Physical Disabilities in South Africa: Beliefs of Non-Disabled People." *Reproductive Health Matters* 25 (50): 66–79. https://doi.org/10.1080/09688080.2017.1332949.

Iltis, Ana S. 2016. "Prenatal Screening and Prenatal Diagnosis: Contemporary Practices in Light of the Past." *Journal of Medical Ethics* 42 (6): 334–39.

Jacobson, Denise Sherer and Anne Finger. 2007. "Alternative Motherhoods." In *Unruly Bodies: Life Writing by Women with Disabilities*, edited by Susannah B. Mintz, 137–82. Chapel Hill, University of North Carolina Press.

Al-Jader, Layla N., and Sharon Hopkins. 2000. "Development of an Antenatal Screening Programme for Congenital Abnormalities in a South Wales District." *Community Genetics* 3 (1): 31–37.

Jaggar, Alison M. 2001. "Is Globalization Good for Women?" *Comparative Literature* 53 (4): 298–314.

———. 2014. "Introduction: Gender and Global Justice: Rethinking Some Basic Assumptions of Western Political Philosophy." In *Gender and Global Justice*, edited by Alison M. Jaggar, 1–17. Malden, MA: Polity Press.

Jennings, Sheila K. 2013. "Reflections on Personhood: Girls with Severe Disabilities and the Law." *Canadian Journal of Disability Studies* 2 (3): 55–97.

Johnson, Harriet McBryde. 2020. "Unspeakable Conversations." In *Disability Visibility: First-Person Stories from the Twenty-first Century*, edited by Alice Wong, 3–27. New York: Vintage Books.

Johnston, Ysabel, Jeffrey P. Bishop, and Griffin Trotter. 2018. "The Moral Imperative to Morally Enhance." *Journal of Medicine and Philosophy* 43 (5): 485–89. https://doi.org/10.1093/jmp/jhy019.

Jonsen, Albert R. 1986. "Bentham in a Box: Technology Assessment and Health Care Allocation." *Law, Medicine and Health Care* 14: 172–74.

Jonsen, Albert R., and Kelly A. Edwards. 2018. "Resource Allocation." UW Medicine. 2018. https://depts.washington.edu/bhdept/ethics-medicine/bioethics-topics/detail/78.

Jordan, Bertrand. 2016. "First Use of CRISPR for Gene Therapy." *Medecine/Sciences* 32 (11): 1035–37.

Jordan, Tina. 2019. "'The Unwinding of the Miracle' Is About How to Die—and Live" *The New York Times,* February 15, 2019. https://www.nytimes.com/2019/02/15/books/review/julie-yip-williams-unwinding-miracle-best-seller.html.

Juengst, Eric T. 1997. "Can Enhancement Be Distinguished from Prevention in Genetic Medicine?" *Journal of Medicine and Philosophy* 22 (2): 125–42. https://doi.org/10.1093/jmp/22.2.125.

———. 2017. "Crowdsourcing the Moral Limits of Human Gene Editing?" *Hastings Center Report* 47 (3): 15–23. https://doi.org/10.1002/hast.701.

Kafer, Alison. 2005. "Hiking Boots and Wheelchairs: Ecofeminism, the Body, and Physical Disability." In *Feminist Interventions in Ethics and Politics: Feminist Ethics and Social Theory*, edited by Barbara S. Andrew, Jean Clare Keller, and Lisa H. Schwartzman, 131–50. New York: Rowman & Littlefield Publishers.

———. 2013. *Feminist, Queer, Crip*. Bloomington: Indiana University Press.

Kamm, Frances M. 2005. "Is There a Problem with Enhancement?" *American Journal of Bioethics* 5 (3): 5–14. https://doi.org/10.1080/15265160590945101.

Kass, Leon R. 2001. "L'Chaim and Its Limits: Why Not Immortality?" *First Things*, 17–24.

———. 2003. *Beyond Therapy: Biotechnology and the Pursuit of Happiness. President's Council on Bioethics.*

Kelly, S. E., and H. R. Farrimond. 2012. "Non-Invasive Prenatal Genetic Testing." *Public Health Genomics* 15 (2): 73–81. https://doi.org/10.1159/000331254.

Khan, Sikandar Hayat. 2019. "Genome-Editing Technologies: Concept, Pros, and Cons of Various Genome-Editing Techniques and Bioethical Concerns for Clinical Application." *Molecular Therapy—Nucleic Acids* 16 (June): 326–34.

Kim, Eunjung. 2011. "'Heaven for Disabled People': Nationalism and International Human Rights Imagery." *Disability & Society* 26(1): 93–106.

———. 2014. "The Spectre of Vulnerability and Disabled Bodies in Protest." in *Disability, Human Rights and the Limits of Humanitarianism*, ed. Michael Gill and Cathy Schlund-Vials, 137–154. Farnham: Ashgate.

Kittay, Eva Feder. 1999. *Love's Labor: Essays on Women, Equality, and Dependency*. New York: Routledge.

———. 2000. “At Home with My Daughter.” In *Americans with Disabilities: Exploring Implications of the Law for Individuals and Institutions,* edited by Leslie P. Francis and Anita Silvers, 64–80. New York: Routledge.

———. 2002a. “Caring for the Vulnerable by Caring for the Caregiver: The Case of Mental Retardation.” In *Medicine and Social Justice: Essays on the Distribution of Health Care*, edited by Anita Silvers, M. Pabst Battin, and Rosamond Rhodes, 290–300. Oxford: Oxford University Press.

———. 2002b. “When Caring Is Just and Justice Is Caring: Justice and Mental Retardation.” In *The Subject of Care: Feminist Perspectives on Dependency*, edited by Eva Feder Kittay and Ellen K. Feder, 257–76. New York: Rowman & Littlefield Publishers.

———. 2006. “The Concept of Care Ethics in Biomedicine: The Case of Disability.” In *Bioethics in Cultural Contexts*, edited by C. Rehmann-Sutter et al., 319–39. Dordrecht, Netherlands: Springer. https://doi.org/10.1007/1-4020-4241-8_22.

———. 2009. “The Personal Is Philosophical Is Political: A Philosopher and Mother of a Cognitively Disabled Person Sends Notes from the Battlefield.” *Metaphilosophy* 40 (3): 606–27.

———. 2011. “The Ethics of Care, Dependence, and Disability.” *Ratio Juris* 24 (1): 49–58. https://doi.org/10.1111/j.1467-9337.2010.00473.x.

———. 2019. *Learning from My Daughter: The Value and Care of Disabled Minds*. New York: Oxford University Press.

Kittay, Eva, Alexa Schriempf, Anita Silvers, and Susan Wendell. 2001. “Introduction.” *Hypatia* 16 (4): vii–xii.

Koch, Tom. 2000. “Life Quality vs the `Quality of Life’: Assumptions Underlying Prospective Quality of Life Instruments in Health Care Planning.” *Social Science & Medicine* 51: 419–27.

Kola´r˘ova, Kater˘ina, and Wiedlack, M. Katharina. 2016. “Introduction.” *Somatechnics, Special Issue: Crip Notes on the Idea of Development* 6 (2): 125–41. https://doi.org/10.3366/soma.2016.0187.

Koo, Taeyoung, Jungjoon Lee, and Jin-soo Kim. 2015. “Measuring and Reducing Off-Target Activities of Programmable Nucleases Including CRISPR-Cas9. MC.Pdf.” *Molecular Cells* 38 (6): 475–81.

Kuppers, Petra. 2014. “Crip Time.” *Tikkun* 29(4): 29–31. https://doi.org/10.1215/08879982-2810062.

La Caze, Marguerite. 2008. “Seeing Oneself through the Eyes of the Other: Asymmetrical Reciprocity and Self-Respect.” *Hypatia: In Honor of Iris Marion Young: Theorist and Practitioner* 23 (3): 118–35. https://doi.org/10.2979/hyp.2008.23.3.118.

Lackey, Jennifer. 2020. “False Confessions and Testimonial Injustice.” *Journal of Criminal Law & Criminology* 110 (1): 43–67.

Lanier, Heather. 2020. *Raising a Rare Girl: A Memoir*. New York: Penguin Random House.

Levinas, Emmanuel. 1998. “Useless Suffering.” In *The Provocation of Levinas: Rethinking the Other*, edited by R. Bernasconi and D. Wood, 156-67. London: Routledge.

Liao, S. Matthew. 2009. “The Right of Children to Be Loved.” In *What Is Right for Children?: The Competing Paradigms of Religion and Human Rights*, edited by Martha Albertson Fineman and Karen Worthington, 347–63.

———. 2010. “The Basis of Human Moral Status.” *Journal of Moral Philosophy* 7: 159–179. https://doi.org/10.1163/174552409X12567397529106.

Lin, Patrick, and Fritz Allhoff. 2006. “Nanoethics and Human Enhancement: A Critical Evaluation of Recent Arguments.” *Nanotechnology Perceptions* 2: 47–52.

Lindemann, Kate. 2001. “Persons with Adult-Onset Head Injury: A Crucial Resource for Feminist Philosophers.” *Hypatia* 16 (4): 105–23. https://doi.org/10.2979/hyp.2001.16.4.105.

Lipsman, Nir, and Walter Glannon. 2013. “Brain, Mind and Machine: What Are the Implications of Deep Brain Stimulation for Perceptions of Personal Agency, Personal Identity, and Free Will?” *Bioethics* 27 (9): 465–70. https://doi.org/10.1111/j.1467-8519.2012.01978.x.

Ljuslinder, Karen, Ellis, Katie and Vikström, Lotta. 2020. Cripping Time—Understanding the Life Course through the Lens of Ableism. *Scandinavian Journal of Disability Research* 22 (1): 35–38. http://doi.org/10.16993/sjdr.710.

Longino, Helen. 2001. *The Fate of Knowledge*. Princeton: Princeton University Press.

Lorde, Audre. 1980. *The Cancer Journals*. San Francisco: Spinsters, Ink.

———. 1988/2017. *A Burst of Light and Other Essays*. Ithaca: Firebrand Books; republished unabridged by Ixia Press.

Mackenzie, Catherine. 2014. “Three Dimensions of Autonomy. A Relational Analysis.” In *Autonomy, Oppression and Gender,* edited by Veltman, A. and M. Piper, 15–42. New York: Oxford University Press.

———. 2015. “Responding to the Agency Dilemma: Autonomy, Adaptive Preferences, and Internalized Oppression.” In *Personal Autonomy and Social Oppression,* edited by Marina Oshana, 48–67. New York: Routledge.

Mackenzie, C. and N. Stoljar, eds). 2000. *Relational Autonomy Feminist Perspectives on Autonomy, Agency and the Social Self.* New York: Oxford University Press.

Malek, Janet. 2010. “Deciding against Disability: Does the Use of Reproductive Genetic Technologies Express Disvalue for People with Disabilities?” *Journal of Medical Ethics* 36 (4): 217–21.

Marcuse, Herbert. 1969. *An Essay on Liberation*. Boston: Beacon Press.

Markens, Susan. 2013. “‘Is This Something You Want?’: Genetic Counselors’ Accounts of Their Role in Prenatal Decision Making.” *Sociological Forum* 28 (3): 431–51.

Marquis, Don. 1989. “Why Abortion Is Immoral.” *The Journal of Philosophy* 86 (4): 183–202.

———. 1995. “Fetuses, Futures, and Values: A Reply to Shirley.” *Southwest Philosophy Review* 6: 263–5.

Marx, Karl.1887. *Capital: A Critique of Political Economy*. Trans. by Samuel Moore and Edward Aveling. Vol. 1. Moscow: Progress Publishers.

———.1932. *Economic and Philosophic Manuscripts of 1844.* Trans. By Martin Milligan. Moscow: Progress Publishers

Mason, Jeff. 2011. "Death and Its Concept." *Philosophersmag.com*. January 31, 2015. https://www.philosophersmag.com/opinion/17-death-and-its-concept.

———. 2012. "Close Encounters of the Cancer Kind." *Philosophersmag.com*. February 1, 2015. https://www.philosophersmag.com/opinion/18-close-encounters-of-the-cancer-kind.

May, Todd. 2017. *A Fragile Life: Accepting Our Vulnerability*. Chicago: University of Chicago Press.

McGruder, Juli. 2004. "Madness in Zanzibar: An Exploration of Lived Experience." In *Schizophrenia, Culture, and Subjectivity: The Edge of Experience*, edited by J.H. Jenkins and R.J. Barrett, 255–281. New York: Cambridge University Press.

McLeod, Carol. 2002. *Self-Trust and Reproductive Autonomy*. Cambridge, MA: MIT Press.

McMahan, Jeff. 1996. "Cognitive Disability, Misfortune, and Justice." *Philosophy & Public Affairs* 25 (1): 3–35. https://doi.org/10.1111/j.1088-4963.1996.tb00074.x.

———. 2002. *The Ethics of Killing: Problems at the Margins of Life*, New York: Oxford University Press.

———. 2005. "Causing Disabled People to Exist and Causing People to be Disabled," *Ethics* 116 (1): 77–99.

———. 2009. "Cognitive Disability and Cognitive Enhancement." *Metaphilosophy* 40: 582–605.

McRuer, Robert. 2006. *Crip Theory: Cultural Signs of Queerness and Disability*. New York: New York University Press.

McWhorter, Ladelle. 2009. "Governmentality, Biopower, and the Debate over Genetic Enhancement." *Journal of Medicine and Philosophy* 34 (4): 409–37. https://doi.org/10.1093/jmp/jhp031.

Meekosha, Helen. 2010. "The Complex Balancing Act of Choice, Autonomy, Valued Life, and Rights: Bringing a Feminist Disability Perspective to Bioethics." *International Journal of Feminist Approaches to Bioethics* 3 (2): 1–8.

Mehlman, Maxwell J. 1999. "How Will We Regulate Genetic Enhancement?" *Wake Forest Law Review* 34 (3): 671–714.

———. 2003. *Wondergenes: Genetic Enhancement and the Future of Society*, Bloomington: Indiana University Press.

———. 2009. *The Price of Perfection: Individualism and Society in the Era of Biomedical Enhancement*. Baltimore: Johns Hopkins University Press.

———. 2012. *Transhumanist Dreams and Dystopian Nightmares*. Baltimore: Johns Hopkins University Press.

Mehlman, Maxwell J., Jessica W. Berg, Eric T. Juengst, and Eric Kodish. 2011. "Ethical and Legal Issues in Enhancement Research on Human Subjects." *Cambridge Quarterly of Healthcare Ethics* 20 (1): 30–45. https://doi.org/10.1017/S0963180110000605.

Menzel, Paul, Paul Dolan, Jeff Richardson, and Jan Abel Abel. 2002. "The Role of Adaptation to Disability and Disease in Health State Valuation: A Preliminary Normative Analysis." *Social Science & Medicine* 55: 2149–58.

Merleau-Ponty, Maurice. 2008. *The Phenomenology of Perception*. Tran. by C. Smith. New York: Routledge.

Meyers, Diana T. 1987. "Personal Autonomy and the Paradox of Feminine Socialization," *Journal of Philosophy*, 84: 619–28.

———. 2005. "Decentralizing Autonomy. Five Faces of Selfhood." In *Autonomy and the Challenges of Liberalism: New Essays,* edited by J. Anderson and J. Christman, 27–55. Cambridge: Cambridge University Press.

Mill, John S. 1869/1998. *On Liberty & Other Essays* (2nd ed.). Edited by J. Gray. New York: Oxford University Press.

———. 1861/2001. *Utilitarianism and the 1868 Speech on Capital Punishment.* Edited by George Sher. Indianapolis: Hackett Publishing.

Mills, Charles W. 2003. *From Class to Race: Essays in White Marxism and Black Radicalism*. Lanham: Rowman and Littlefield.

———. 2017. *Black Rights/White Wrongs: The Critique of Racial Liberalism*. New York: Oxford University Press.

Mitchell, David T., and Sharon L. Snyder. 2016. "The Matter of Disability." *Bioethical Inquiry* 13: 487–92.

Moser, Ingunn. 2006. "Sociotechnical Practices and Difference: On the Interferences between Disability, Gender, and Class." *Science Technology and Human Values*. https://doi.org/10.1177/0162243906289611.

Nelkin, Dorothy and M. Susan Lindee. 2004. *The DNA Mystique: The Gene as a Cultural Icon.* Ann Arbor: University of Michigan Press.

Nelson, M. K., Shew, A., and Stevens, B. 2019. "Transmobility: Rethinking the Possibilities in Cyborg (Cripborg) Bodies." *Catalyst: Feminism, Theory, Technoscience* 5 (1): 1–20.

Nicki, Andrea. 2001. "The Abused Mind: Feminist Theory, Psychiatric Disability, and Trauma." *Hypatia* 16 (4): 80–104. https://doi.org/10.2979/hyp.2001.16.4.80.

Nocella, Anthony J., II, Judy K. C. Bentley, and Janet M. Duncan, ed. 2012. *Earth, Animal, and Disability Liberation: The Rise of the Eco-Ability Movement*. New York: Peter Lang Publishing.

Nocella, Anthony J., II, Amber E. George, and John Lupinacci. 2019. *Animals, Disability, and the End of Capitalism: Voices from the Eco-ability Movement*. New York: Peter Lang Publishing.

Nussbaum, Martha C. 2001. "The Enduring Significance of John Rawls." *The Chronicle of Higher Education* 47 (45): 1–7.

———. 2006. *Frontiers of Justice: Disability, Nationality, Species Membership*. Cambridge: Harvard University Press.

———. 2009. "The Capabilities of People with Cognitive Disabilities." *Metaphilosophy* 40 (3): 331–51.

Orentlicher, David. 1996. "Destructuring Disability: Rationing of Health Care and Unfair Discrimination against the Sick." *Harvard Civil Rights-Civil Liberties Law Review* 31: 49–89.

Oshana, Marina. 2006. *Personal Autonomy in Society*. Aldershot: Ashgate Publishing.

Oskrochi, G., Ahmed Bani-Mustafa, and Y. Oskrochi. 2018. "Factors Affecting Psychological Well-Being: Evidence from Two Nationally Representative Surveys." *PLoS ONE* 13 (6): 1–14. https://doi.org/10.1371/journal.pone.0198638.

Parens, Erik. 1995. "The Goodness of Fragility: On the Prospect of Genetic Technologies Aimed at the Enhancement of Human Capacities." *Kennedy Institute of Ethics Journal* 5 (2): 141–53. https://doi.org/10.1353/ken.0.0149.

———. 2005. "Authenticity and Ambivalence: Toward Understanding the Enhancement Debate." *Hastings Center Report* 35 (3): 34–41.

———, ed. 2006. *Surgically Shaping Children: Technology, Ethics, and the Pursuit of Normality*. Baltimore: Johns Hopkins University Press.

———. 2013. "On Good and Bad Forms of Medicalization." *Bioethics* 27 (1): 28–35. https://doi.org/10.1111/j.1467-8519.2011.01885.x.

Parens, Erik, and Adrienne Asch. 1999. "The Disability Rights Critique of Prenatal Genetic Testing: Reflections and Recommendations." *Hastings Center Report* 29 (5): S1–S22.

Paul, Diane B. 1995. *Controlling Human Heredity: 1865 to the Present*. Atlantic Highlands, NJ: Humanities Press.

Pavone, Vincenzo, and Flor Arias. 2012. "Geneticization Thesis: The Beyond the Political Economy of PGD/PGS in Spain." *Science, Technology, & Human Values* 37 (3): 235–61. https://doi.org/10.1177/016224391.

Penchaszadeh, Victor B., and Diana Puñales-Morejón. 1998. "Genetic Services to the Latino Population in the United States." *Community Genetics* 1 (3): 134–41.

Persson, Ingmar, and Julian Savulescu. 2008. "The Perils of Cognitive Enhancement and the Urgent Imperative to Enhance the Moral Character of Humanity." *Journal of Applied Philosophy* 25 (3): 162–77. https://doi.org/https://doi.org/10.1111/j.1468-5930.2008.00410.x.

———. 2012. *Unfit for the Future: The Need for Moral Enhancement*. Oxford: Oxford University Press. https://doi.org/10.1093/analys/ant021.

———. 2014. "Against Fetishism about Egalitarianism and in Defense of Cautious Moral Bioenhancement." *American Journal of Bioethics* 14 (4): 39–42. https://doi.org/10.1080/15265161.2014.889248.

———. 2015. "The Art of Misunderstanding Moral Bioenhancement." *Cambridge Quarterly of Healthcare Ethics* 24 (1): 48–57.

Peters, Philip G., Jr. 1995. "Health Care Rationing and Disability Rights." *Indiana Law Journal* 70 (2): 491–548.

Price, Margaret. 2011. *Mad at School: Rhetorics of Mental Disability and Academic Life*. Ann Arbor: University of Michigan Press.

Purcell, Elizabeth. 2014. "Oppression's Three New Faces: Rethinking Iris Young's 'Five Faces of Oppression' for Disability Theory." In *Diversity, Social Justice and Inclusive Excellence: Transdisciplinary and Global Perspectives*, edited by Nagel Mecke and Seth Asumah, 185–205. Albany: State University of New York Press.

———. 2016a. "Ethics and Mental Health: An Intercultural Approach." *Wagadu: A Journal of Transnational Women's and Gender Studies* 15 (Special Issue: Epistemic Injustice in Practice): 115–38.

———. 2016b. "Disability, Narrative, and Moral Status." *Disability Studies Quarterly* (36) 1.

———. 2019. "Iris Marion Young's City of Difference." In *Handbook of Philosophy of the City*, edited by Sharon M. Meagher, Samantha Noll, and Joseph S. Biel, 101-10. New York: Routledge Press.

Purdy, Laura M. 1995. "Loving Future People." In *Reproduction, Ethics, and the Law: Feminist Perspectives, 300–27.* Bloomington: Indiana University Press.

Race, D., K. Boxall, and I. Carson. 2005. "Towards a Dialogue for Practice: Reconciling Social Role Valorization and the Social Model of Disability." In *Disability and Society* 20 (5): 507–21.

Rakić, Vojin. 2015. "We Must Create Beings with Moral Standing Superior to Our Own." *Cambridge Quarterly of Healthcare Ethics* 24 (1): 58–65. https://doi.org/10.1017/S0963180114000309.

Rawls, John. 1971. *A Theory of Justice*. Cambridge: Harvard University Press.

———. 1985. "Justice as Fairness: Political Not Metaphysical." *Philosophy & Public Affairs* 14 (3): 223–51.

Reeve, Donna. 2009. "Revisiting the 'Dys-Appearing Body' through the Lens of Psycho-Emotional Disablism." Paper presented at the *9th Conference of European Sociological Association*, Lisbon: 2–5.

Resnik, David B., and Daniel B Vorhaus. 2006. "Genetic Modification and Genetic Determinism." *Philosophy, Ethics, and Humanities Genetic Modification and Genetic Determinism* 1 (9): 1–11. https://doi.org/10.1186/1747-5341-1-9.

Rice, C., E. Chandler, J. Rinaldi, N. Changfoot, K. Liddiard, R. Mykitiuk,, and I. Mündel. 2017. "Imagining Disability Futurities." *Hypatia 32* (2): 213–29. doi:10.1111/hypa.12321.

Ricoeur, Paul. 1992. *Oneself as Another*. Trans. by K. Blamey. Chicago, IL: University of Chicago Press.

Riddle, Christopher A. 2014a. "Disability and Justice." In *Disability and Justice: The Capabilities Approach in Practice*, 1–6. New York: Lexington Books. https://doi.org/10.1080/15017419.2015.1064029.

———. 2014b. "The Special Moral Importance of Health." In *Disability and Justice: The Capabilities Approach in Practice*, 31–46. New York: Lexington Books. https://doi.org/10.1057/978-1-137-59993-3_4.

Robert, Jason Scott, and Françoise Baylis. 2003. "Crossing Species Boundaries." *American Journal of Bioethics* 3 (3): 1–13. https://doi.org/10.1162/15265160360706417.

Roberts, Dorothy E. 2009. "Race, Gender, and Genetic Technologies: A New Reproductive Dystopia?" *Signs* 34 (4): 783–804.

Robinson, Cedric J. 1983/2000. *Black Marxism: The Making of the Black Radical Tradition*. Chapel Hill: University of North Carolina Press.

———. 2019. *On Racial Capitalism, Black Internationalism, and Cultures of Resistance*. Edited by H.L.T. Quan, 221–32. London: Pluto Press.

Rock, Melanie. 2000. "Discounted Lives? Weighing Disability When Measuring Health and Ruling on 'Compassionate' Murder." *Social Science & Medicine* 51: 407–17.

Roduit, Johann A. R., Holger Baumann, and Jan-Christoph Heilinger. 2013. "Human Enhancement and Perfection." *Journal of Medical Ethics* 39 (10): 647–50.

Rohwerder, Brigitte. 2018. *Disability Stigma in Developing Countries*. K4D Helpdesk Report. Brighton, UK: Institute of Development Studies.

Ross, Catherine E. and Marieke van Willigen. 1997. "Education and the Subjective Quality of Life." *Journal of Health and Social Behavior* 38: 275–97.

Routtenberg, A., I. Cantallops, S. Zaffuto, P. Serrano, and U. Namgung. 2000. "Enhanced Learning after Genetic Overexpression of a Brain Growth Protein." Proceedings of the National Academy of Sciences of the United States of America, 97 (13): 7657–62.

Russell, Marta. 2019. *Capitalism and Disability*, edited by Keith Rosenthal. Chicago: Haymarket Books.

Samuels, Ellen. 2017. "Six Ways of Looking at Crip Time." *Disability Studies Quarterly* 37 (3). https://doi. org/10.18061/dsq.v37i3.5824

Sandahl, Carrie. 2003. "Queering the Crip or Cripping the Queer? Intersections of Queer and Crip Identities in Solo Autobiographical Performance." *GLQ: A Journal of Lesbian and Gay Studies* 9 (1–2): 25–56.

Sandel, Michael J. 2007. *The Case Against Perfection: Ethics in the Age of Genetic Engineering*. Cambridge: Harvard University Press.

Sanders, Robert. 2021. "FDA Approves First Test of CRISPR to Correct Genetic Defect Causing Sickle Cell Disease." *Berkeley News,* March 30. https://news.berkeley.edu/2021/03/30/fda-approves-first-test-of-crispr-to-correct-genetic-defect-causing-sickle-cell-disease/.

Sarton, May. 2007. "A Brief History of Aging." In *Unruly Bodies: Life Writing by Women with Disabilities*, edited by Susannah B. Mintz, 183–210. Chapel Hill, University of North Carolina Press.

Savulescu, Julian. 2001. "Procreative beneficence: why we should select the best children." *Bioethics* 15. (5-6): 413-26. doi:10.1111/1467-8519.00251.

———. 2006. "Justice, Fairness, and Enhancement." *Annals of the New York Academy of Sciences* 1093: 321–38. https://doi.org/10.1196/annals.1382.021.

———. 2007. "In Defence of Procreative Beneficience." *Journal of Medical Ethics* 33: 284–88.

———. 2009a. "Autonomy, Well-Being, Disease, and Disability." *PPP* 16 (1): 58–65.

———. 2009b. "The Human Prejudice and the Moral Status of Enhanced Beings: What Do We Owe the Gods?" In *Human Enhancement*, edited by J. Savulescu and N. Bostrom, 211–50. New York: Oxford University Press.

Schermer, Maartje. 2008. "Enhancements, Easy Shortcuts, and the Richness of Human Activities." *Bioethics* 22 (7): 355–63. https://doi.org/10.1111/j.1467-8519.2008.00657.x.

Schriempf, Alexa. 2001. "(Re)Fusing the Amputated Body: An Interactionist Bridge for Feminism and Disability." *Hypatia* 16 (4): 53–79. https://doi.org/10.2979/hyp.2001.16.4.53.

Scott, Rosamund. 2006. "Choosing between Possible Lives: Legal and Ethical Issues in Preimplantation Genetic Diagnosis." *Oxford Journal of Legal Studies* 26 (1): 153–78.

Scullion, Philip Andrew. 2010. "Models of Disability: Their Influence in Nursing and Potential Role in Challenging Discrimination." *Journal of Advanced Nursing*. https://doi.org/10.1111/j.1365-2648.2009.05211.x.

Scuro, Jennifer. 2018. *Addressing Ableism: Philosophical Questions via Disability Studies*. Lanham: Lexington Books.

Segal, Teresa M. 2010. "The Role of the Reproductive Technology Clinic in the Imposition of Societal Values." *International Journal of Feminist Approaches to Bioethics* 3 (2): 90–108.

Selgelid, Michael J. 2014. "Moderate Eugenics and Human Enhancement." *Medicine, Health Care and Philosophy* 17 (1): 3–12. https://doi.org/10.1007/s11019-013-9485-1.

Shakespeare, Torn, and Nicholas Watson. 2001. "The Social Model of Disability: An Outdated Ideology?" *Research in Social Science and Disability*. https://doi.org/10.1016/S1479-3547(01)80018-X.

Shanley, Mary Lyndon, and Adrienne Asch. 2009. "Involuntary Childlessness, Reproductive Technology, and Social Justice: The Medical Mask on Social Illness." *Signs: Journal of Women in Culture and Society* 34 (4): 851–74. https://doi.org/10.1086/597141.

Shapiro, J. P. 1994. *No Pity: People with Disabilities Forging a New Civil Rights Movement*. New York: Three Rivers.

Shelby, Tommie. 2005. *We Who Are Dark: The Philosophical Foundations of Black Solidarity*. Cambridge: Harvard University Press.

———. 2016. *Dark Ghettos: Injustice, Dissent, and Reform*. Cambridge: Harvard University Press.

Shickle, Darren. 2000. "Are 'Genetic Enhancements' Really Enhancements?" *Cambridge Quarterly of Healthcare Ethics* 9 (3): 342–52. https://doi.org/10.1017/s0963180100903062.

Sikela, James M. 2006. "The Jewels of Our Genome: The Search for the Genomic Changes Underlying the Evolutionarily Unique Capacities of the Human Brain." *PLoS Genetics* 2 (5): 646–55.

Silberman, Steve. 2015. *NeuroTribes: The Legacy of Autism and the Future of Neurodiversity*. New York: Penguin Random House.

Silvers, Anita. 1995. "Reconciling Equality to Difference: Caring (F) or Justice For People With Disabilities." *Hypatia* 10 (1): 30–55. https://doi.org/10.1111/j.1527-2001.1995.tb01352.x.

———. 1998. "A Fatal Attraction to Normalizing: Treating Disabilities as Deviations from "'Species-Typical' Functioning." In *Enhancing Human Traits*, edited by E. Parens, 177–202. Washington, DC: Georgetown University Press.

———. 2007. "Liberalism and Individually Scripted Ideas of the Good: Meeting the Challenge of Dependent Agency." *Social Theory and Practice* 33 (2): 311–33.

———. 2012. "Moral Status: What a Bad Idea!" *Journal of Intellectual Disability Research* 56 (11): 1014–25. https://doi.org/10.1111/j.1365-2788.2012.01616.x.

Silvers, Anita, and Leslie Pickering Francis. 2005. "Justice through Trust: Disability and the 'Outlier Problem' in Social Contract Theory." *Ethics*. https://doi.org/10.1086/454368.

———. 2010. “Thinking about the Good: Reconfiguring Liberal Metaphysics (or Not) for People with Cognitive Disabilities.” In *Cognitive Disability and Its Challenge to Moral Philosophy*, 237–59. https://doi.org/10.1017/CBO9781107415324.004.

Silvers, Anita, and M. A. Stein. 2002. “Disability, Equal Protection, and the Supreme Court: Standing at the Crossroads of Progressive and Retrogressive Logic in Constitutional Classification.” *The Michigan Journal of Law Reform*, 35 (1 and 2): 81.

———.2003. “Human Rights and Genetic Discrimination: Protecting Genomics’ Promise for Public Health.” *Journal of Medicine, Law and Ethics* 31 (3): 377–89.

Singer, Peter. 2008 “Parental Choice and Human Improvement.” In *Human Enhancement*, edited by Julian Savulescu and Nick Bostrom, 277–90. New York: Oxford University Press.

———. 2009. “Speciesism and Moral Status.” *Metaphilosophy* 40: 567–81. https://doi.org/10.1002/9781444322781.ch19.

Slorach, Roddy. 2016. *A Very Capitalist Condition: A History and Politics of Disability*. London: Bookmarks Publications.

Solomon, Miriam. 2015. *Making Medical Knowledge*. Oxford: Oxford University Press.

Sorgner, Stefan Lorenz. 2015. “The Future of Education: Genetic Enhancement and Metahumanities.” *Journal of Evolution and Technology* 25 (May 1): 31–48.

Sparrow, Robert. 2011. “A Not-So-New Eugenics: Harris and Savulescu on Human Enhancement.” *The Hastings Center Report* 41 (1): 32–42.

———. 2014a. “Better Living Through Chemistry? A Reply to Savulescu and Persson on ‘Moral Enhancement.’” *Journal of Applied Philosophy* 31 (1): 23–32. https://doi.org/10.1111/japp.12038.

———. 2014b. “Egalitarianism and Moral Bioenhancement.” *American Journal of Bioethics* 14 (4): 20–28. https://doi.org/10.1080/15265161.2014.889241.

———. 2015. “Enhancement and Obsolescence: Avoiding an ‘Enhanced Rat Race.’” *Kennedy Institute of Ethics Journal* 25 (3): 231–60. https://doi.org/10.1353/ken.2015.0015.

Specker, Jona, and Maartje H.N. Schermer. 2017. “Imagining Moral Bioenhancement Practices: Drawing Inspiration from Moral Education, Public Health Ethics, and Forensic Psychiatry.” *Cambridge Quarterly of Healthcare Ethics* 26 (3): 415–26. https://doi.org/10.1017/S0963180116001080.

Stein, Mark S. 2006. *Distributive Justice and Disability: Utilitarianism against Egalitarianism*. New Haven: Yale University Press.

Steinbock, Bonnie. 2011. *Life Before Birth: The Moral and Legal Status of Embryos and Fetuses.* 2nd ed. New York: Oxford University Press.

Stramondo, Joseph A. 2011. “Doing Ethics from Experience: Pragmatic Suggestions for a Feminist Disability Advocate’s Response to Prenatal Diagnosis.” *International Journal of Feminist Approaches to Bioethics* 4 (2): 48–78.

Tonkens, Ryan. 2011. “Parental Wisdom, Empirical Blindness, and Normative Evaluation of Prenatal Genetic Enhancement.” *Journal of Medicine and Philosophy* 36: 274–95. https://doi.org/10.1093/jmp/jhr012.

Tremain, Shelley L., ed. 2015. *Foucault and the Government of Disability*. Ann Arbor: University of Michigan Press.

———. 2017. *Foucault and Feminist Philosophy of Disability*. Ann Arbor: University of Michigan Press.

Tuana, Nancy, and Shannon Sullivan. 2006. "Introduction: Feminist Epistemologies of Ignorance." *Hypatia* 21 (3): vii–ix. https://doi.org/10.1353/hyp.2006.0036.

Ubel, Peter A., George Loewenstein, and Christopher Jepson. 2003. "Whose Quality of Life? A Commentary Exploring Discrepancies between Health State Evaluations of Patients and the General Public." *Quality of Life Research* 12 (6): 599–607.

van Hooft, Stan. 1998. "The Meanings of Suffering," *Hastings Center Report* 28 (5): 13–19.

Veit, Walter. 2018. "Procreative Beneficence and Genetic Enhancement." *Kriterion—Journal of Philosophy*, 1–18.

Vig, Elizabeth K., and Pearlman Robert A. 2004. "Good and Bad Dying From the Perspective of Terminally Ill Men." *Archives of Internal Medicine*. 164 (9): 977–81. doi:10.1001/archinte.164.9.977.

Wachbroit, R. and D. Wasserman. 2003. "Reproductive Technology." In *The Oxford Handbook of Practical Ethics*, edited by Hugh Lafollette, 136–160. New York: Oxford University Press.

Waples, Emily. 2014. "Avatars, Illness, and Authority: Embodied Experience in Breast Cancer Autopathographics." *Configurations* 22 (2) (Spring): 153–81.

Walsh, Pat. 2010. "Asperger Syndrome and the Supposed Obligation Not to Bring Disabled Lives into the World." *Journal of Medical Ethics*. https://doi.org/10.1136/jme.2010.036459.

Wang, H. B., G. D. Ferguson, V. V. Pineda, P. E. Cundiff, and D. R. Storm. 2004. "Overexpression of Type-1 Adenylyl Cyclase in Mouse Forebrain Enhances Recognition Memory and LTP." *Nature Neuroscience*, 7(6): 635–42.

Wasserman, David. 1998. "Distributive Justice," in *Disability, Difference, Discrimination: Perspectives on Justice in Bioethics and Public Policy*, edited by Anita Silvers, David T. Wasserman, and Mary Briody Mahowald, 147–208. Lanham, MD: Rowman and Littlefield.

Wasserman, David, and Adrienne Asch. 2012a. "A Duty to Discriminate?" *American Journal of Bioethics* 12 (4): 22–24. https://doi.org/10.1080/15265161.2012.656814.

———. 2012b. "Reproductive Medicine and Turner Syndrome: Ethical Issues." *Fertility and Sterility* 98 (4): 792–96. https://doi.org/10.1016/j.fertnstert.2012.08.036.

Wasserman, David, Jerome Bickenbach, and Robert Wachbroit, eds. 2005. *The Quality of Life and Human Difference: Genetic Testing, Health Care, and Disability*. New York: Cambridge University Press.

Watters, Ethan. 2010. *Crazy Like Us: The Globalization of the American Psyche*. New York: Free Press.

Wei, F., G. D. Wang, G. A. Kerchner, S. J. Kim, H. M. Xu, Z. F. Chen, and M. Zhuo. 2001. "Genetic Enhancement of Inflammatory Pain by Forebrain NR2B Overexpression." *Nature Neuroscience*, 4 (2): 164–69.

Weil, Simone. 1977. "The Love of God and Affliction." In *The Simone Weil Reader*, edited by George A. Panichas , 31-45. New York: David McKay Company.

Weir, Allison. 2008. "Home and Identity: In Memory of Iris Marion Young." *Hypatia: A Journal of Feminist Philosophy* 23 (3): 4–21. https://doi.org/10.2979/hyp.2008.23.3.4.

Wendell, Susan. 1989. "Toward a Feminist Theory of Disability." *Hypatia* 4 (2): 104–24. https://doi.org/10.4324/9781315189413.

———. 1996. *The Rejected Body: Feminist Philosophical Reflections on Disability*. New York: Routledge.

———. 2016. "Unhealthy Disabled: Treating Chronic Illnesses as Disabilities." *The Disability Studies Reader. 5th ed.* https://doi.org/10.4324/9781315680668.

Wertz, D. 1998. "What's Missing from Genetic Counseling: A Survey of 476 Counseling Sessions," abstract, *Journal of Genetic Counseling* 7 (6): 499–500.

West, Michael D., Hal Sternberg, Ivan Labat, Jeffrey Janus, Karen B. Chapman, Nafees N. Malik, Aubrey D.N.J. De Grey, and Dana Larocca. 2019. "Toward a Unified Theory of Aging and Regeneration." *Regenerative Medicine* 14 (9): 867–86. https://doi.org/10.2217/rme-2019-0062.

Wikler, Daniel. 1999. "Can We Learn from Eugenics?" *Journal of Medical Ethics* 25 (2): 183–94. https://doi.org/10.1136/jme.25.2.183.

———. 2002. "Personal and Social Responsibility for Health." *Ethics and International Affairs* 16 (2): 47–55.

———. 2009. "Paternalism in the Age of Cognitive Enhancement: Do Civil Liberties Presuppose Roughly Equal Mental Ability?" In *Human Enhancement*, edited by J. Savulescu and N. Bostrom, 341–56. Oxford: Oxford University Press.

Wise, Tim. 2009. *Between Barack and a Hard Place: Racism and White Denial in the Age of Obama*. San Francisco: City Light Publishers.

Wolbring, Gregor. 2012. "Ethical Theories and Discourses through an Ability Expectations and Ableism Lens: The Case of Enhancement and Global Regulation." *Asian Bioethics Review* 4 (4): 293–309.

Wong, Alice, ed. 2020. *Disability Visibility: First-Person Stories from the Twenty-First Century*. New York: Vintage Books.

Wong, Sophia Isako. 2007. "The Moral Personhood of Individuals Labeled 'Mentally Retarded': A Rawlsian Response to Nussbaum." *Social Theory and Practice*, 33 (4): 579–94. doi:10.5840/soctheorpract20073343

———. 2010. "Duties of Justice to Citizens with Cognitive Disabilities." In *Cognitive Disability and Its Challenge to Moral Philosophy*. https://doi.org/10.1002/9781444322781.ch7.

Woodard, Christopher. 2013. "Classifying Theories of Welfare." *Philosophical Studies: An International Journal for Philosophy in the Analytic Tradition* 165 (3): 787–803. https://doi.org/10.1007/sl.

Woodcock, Scott. 2009. "Disability, Diversity, and the Elimination of Human Kinds." *Social Theory and Practice* 35 (2): 251–78.

World Health Organization (WHO). 2011. "Assistive Technology." https://www.who.int/news-room/fact-sheets/detail/assistive-technology

World Health Organization (WHO). 2018. "World Health Statistics 2018: Monitoring Health for Sustainable Development Goals." https://apps.who.int/iris/bitstream/handle/10665/272596/9789241565585-eng.pdf?ua=1

Wu, Katherine J. 2020. "Crispr Gene Editing Can Cause Unwanted Changes in Human Embryos, Study Finds: Instead of Addressing Genetic Mutations, the Crispr Machinery Prompted Cells to Lose Entire Chromosomes." *The New York Times,* October 31.

Yergeau, Melanie. 2018. *Authoring Autism: On Rhetoric and Neurological Queerness*. Durham: Duke University Press.

Yip-Williams, Julie. 2018. "A Woman with Cancer Faces Her End." *CBS Sunday Morning*. March 11, 2018. https://www.youtube.com/watch?v=_hW5Fnwp4kk.

Young, Iris Marion. 1990a. "The Distributive Paradigm." In *Justice and the Politics of Difference*, 15–38. Princeton: Princeton University Press.

———. 1990b. "City Life and Difference." In *Justice and the Politics of Difference*, 250–70. Princeton: Princeton University Press.

———. 1990c. "Faces of Oppression." In *Justice and the Politics of Difference*, 39–65. Princeton: Princeton University Press.

———. 1990d. "Introduction." In *Justice and the Politics of Difference*, 1–14. Princeton: Princeton University Press.

———. 1990e. "Social Movements and the Politics of Difference." *Justice and the Politics of Difference*, 156–83. Princeton: Princeton University Press.

———. 1997. "Asymmetrical Reciprocity: On Moral Respect, Wonder and Enlarged Thought." *Feminism and the Public Sphere* 3 (3): 340–63.

———. 2000. *Inclusion and Democracy*. Oxford: Oxford University Press.

———. 2005. *On Female Body Experience: "Throwing Like a Girl" and Other Essays*. New York: Oxford University Press.

Index

About the Author

Elyse Purcell is an Assistant Professor of Philosophy at the State University of New York College at Oneonta. She has published numerous articles in bioethics, social and political philosophy and the medical humanities. Always an interdisciplinary thinker, she aims to bring the insights from fields such as disability studies into contemporary philosophical discussions and debates. Elyse also serves as the Secretary-Treasurer for the American Philosophical Association Central Division.